An Introduction to the Lattice Boltzmann Method

A Numerical Method for Complex Boundary
and Moving Boundary Flows

An Introduction to the Lattice Boltzmann Method

A Numerical Method for Complex Boundary and Moving Boundary Flows

Takaji Inamuro
Kyoto University, Japan

Masato Yoshino
Shinshu University, Japan

Kosuke Suzuki
Shinshu University, Japan

Ⓜ **MARUZEN**
PUBLISHING

Ⓦ**ⓔ** World Scientific

Published by

World Scientific Publishing Co. Pte. Ltd.
5 Toh Tuck Link, Singapore 596224
USA office: 27 Warren Street, Suite 401-402, Hackensack, NJ 07601
UK office: 57 Shelton Street, Covent Garden, London WC2H 9HE

and

Maruzen Publishing Co., Ltd.
Kanda Jimbo-cho Bldg. 6F, Kanda Jimbo-cho 2-17
Chiyoda-ku, Tokyo 101-0051, Japan

Library of Congress Control Number: 2021050556

British Library Cataloguing-in-Publication Data
A catalogue record for this book is available from the British Library.

AN INTRODUCTION TO THE LATTICE BOLTZMANN METHOD
A Numerical Method for Complex Boundary and Moving Boundary Flows

ISBN 978-981-124-051-5 (hardcover)
ISBN 978-981-124-052-2 (ebook for institutions)
ISBN 978-981-124-053-9 (ebook for individuals)

For any available supplementary material, please visit
https://www.worldscientific.com/worldscibooks/10.1142/12375#t=suppl

Preface

This book presents the fundamentals and applications of the lattice Boltzmann method, one of the numerical methods for incompressible viscous fluid flows.

One of the features of the numerical methods for incompressible viscous fluid flows is how to find the pressure that satisfies the continuity equation because there is no time evolution equation for finding the pressure. Looking at the history of this topic, numerical methods for incompressible viscous fluid flows have been developed since the 1960s, as represented by the MAC method, and the basic scheme was almost completed in the 1980s. In the basic scheme, the pressure satisfying the continuity equation is obtained by solving the Poisson equation of pressure at each time. Finding this solution consumes most of the calculation time. An alternative method, also proposed in the 1960s, is the artificial compressibility method. Instead of solving the Poisson equation of pressure, this method adds a derivative term of pressure by pseudo-time to the continuity equation. The basic scheme of this method including unsteady flow calculations was almost completed in the early 1990s.

Since the 1990s, the lattice Boltzmann method (hereafter referred to as the LBM) has been proposed as a numerical method for incompressible viscous fluid flows. In the LBM, a modeled fluid, which is composed of identical fictitious particles whose velocities are restricted to a finite set of vectors (called a lattice gas model), is considered. The streaming and collision processes of each particle are computed with a particle velocity distribution function, and the macroscopic quantities such as mass density and momentum density are obtained from the moments of the velocity distribution function. In other words, the LBM is a numerical method that uses an analogy to the kinetic theory of gases to study fluid flows from a microscopic standpoint. To understand the LBM, researchers therefore require some knowledge of the kinetic theory of gases, that is, the Boltzmann equation. For this reason (among others), the LBM is regarded as a particular numerical scheme that is difficult to understand for researchers familiar with the Navier–Stokes equations. However, once the basic ideas are grasped, the LBM is a very simple and efficient numerical scheme. In particular, it removes the need for solving the Poisson equation of pressure and has a simple algorithm for complicated flow fields. It also has good conservation of mass and momentum and is suitable for parallel computation. Thus far, the LBM has been successfully applied to flow fields with complex boundaries such as flows in porous media, and to flow fields with complicated interfacial changes such as moving boundary flows and gas–liquid two-phase flows. Unfortunately, owing to rapid development, methods are commonly applied without understanding the basic ideas, leading to misevaluations of the LBM.

This book gives the fundamentals and applications of the LBM. Most of the contents are based on the authors' results of previous studies. The first author has also written several reviews of the LBM [61,65,72,73,75], but this book is written for easy understanding of both the reviews and the latest research results. The book is organized as follows. In Chap. 1, we formulate the LBM for incompressible viscous fluid flows and clarify the relationship between the LBM and the Navier–Stokes equations in an asymptotic analysis. We also present a numerical example of complex boundary flows. Chapter 2 introduces the lattice kinetic scheme developed from the LBM. Chapter 3 describes the immersed boundary LBM applicable to moving boundary flows. Chapter 4 describes the two-phase LBM applicable to gas–liquid and liquid–liquid two-phase fluid flows. Regarding a two-phase interface as a boundary, a two-phase fluid flow is one of the moving boundary flows. To assist readers, the contents of each chapter are also summarized in a section on computational algorithms. Appendix F contains the download URL of simple program examples (source codes) and their explanations. Referring to these, readers are encouraged to program the computational algorithms and compute a simple example to appreciate the advantages of the LBM. We hope to inspire interest and further development of the LBM in future studies. We would appreciate readers using this book as a guide to the LBM.

Note that good books and reviews on the LBM have been published elsewhere [2,15,114,130,154]. The Japanese version of this book was published by Maruzen Publishing in January 2020.

Finally, we sincerely thank the people of Maruzen Publishing and World Scientific Publishing who engaged in the publishing of this book. We also thank Crimson Interactive Pvt. Ltd. for the English proofreading of the manuscript.

October, 2021
Authors

Contents

Chapter 1

Lattice Boltzmann Method (LBM)

In the lattice Boltzmann method (hereafter referred to as the LBM), we consider a modeled fluid which is composed of identical fictitious particles whose velocities are restricted to a finite set of vectors (called a lattice gas model). The streaming and collision processes of each particle are computed with a particle velocity distribution function, and macroscopic quantities such as mass density and momentum density are obtained from the moments of the velocity distribution function. In other words, the LBM is a numerical method that uses an analogy to the kinetic theory of gases to study fluid flows from a microscopic standpoint. It should be noted that the LBM formulation (unlike particle methods) does not directly track the behavior of the fictitious particles.

This chapter explains the LBM for a single-phase incompressible viscous fluid.[1] In incompressible fluids, the velocity and pressure fields can be determined separately from the temperature field. That is, the flow velocity u and pressure p can be obtained by the continuity equation and the Navier–Stokes equations, while the fluid temperature T is determined by the advection–diffusion equation in the velocity field u. Therefore, we first describe the LBM for calculating the flow velocity u and pressure p of the isothermal field. We then describe the LBM for obtaining the temperature field.

In the following, we will use the dimensionless variables described in Appendix A.[2] We also define two time scales, the "acoustic time scale" that handles fast phenomena related to movement of the fictitious particles, and the "diffusive time scale" that handles slow phenomena related to the fluid movement. Unless otherwise specified, this book uses the diffusive time scale.

---

[1]As will be seen later, such a fluid is not 'strictly incompressible' but 'almost incompressible (weakly compressible).'

[2]To avoid the repeated use of symbols, we will use the same symbol in a different sense where there is no misunderstanding.

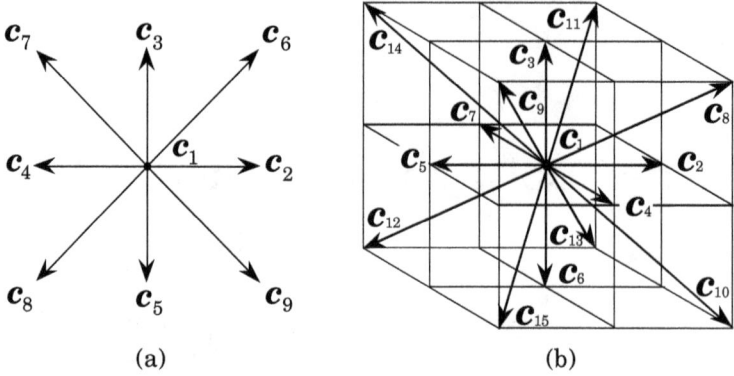

Fig. 1.1   Lattice gas model: (a) D2Q9 model; (b) D3Q15 model.

1.1   Lattice Gas Model

Although several lattice gas models have been proposed for isothermal fields,[3] the formulation in this book adopts a two-dimensional nine-velocity model (called the D2Q9 model) and a three-dimensional fifteen-velocity model (called the D3Q15 model) (see Fig. 1.1). The particle velocities in the two-dimensional nine-velocity model are given by

$$c_i = \begin{cases} (0,0), & i = 1, \\ (\pm 1, 0), (0, \pm 1), & i = 2,3,4,5, \\ (\pm 1, \pm 1), & i = 6,7,8,9. \end{cases} \tag{1.1}$$

In the three-dimensional fifteen-velocity model, the particle velocities are given by

$$c_i = \begin{cases} (0,0,0), & i = 1, \\ (\pm 1, 0, 0), (0, \pm 1, 0), (0, 0, \pm 1), & i = 2,3,\cdots,7, \\ (\pm 1, \pm 1, \pm 1), & i = 8,9,\cdots,15. \end{cases} \tag{1.2}$$

In addition, a three-dimensional nineteen-velocity model (D3Q19 model) and a three-dimensional twenty-seven-velocity model (D3Q27 model) are often used (see Appendix B).

---

[3] These models are also called athermal models.

1.2   Basic Equation

We first write the governing equation of the velocity distribution function $f_i(\boldsymbol{x}, t)$ of a fictitious particle having velocity $\boldsymbol{c}_i$ at coordinates $\boldsymbol{x}$ at time t. The governing equation is the following discrete Boltzmann equation [56]:

$$\mathrm{Sh}\frac{\partial f_i}{\partial t} + \boldsymbol{c}_i \cdot \nabla f_i = \frac{1}{\varepsilon}\Omega_i[\boldsymbol{f}(\boldsymbol{x}, t)], \qquad i = 1, 2, \cdots, N, \qquad (1.3)$$

where $\mathrm{Sh} = \hat{L}/(\hat{t}_0\hat{c}) = \hat{U}/\hat{c}$ is the Strouhal number,[4] ε is a small parameter equivalent to the Knudsen number Kn ($= \hat{\ell}/\hat{L}$, where $\hat{\ell}$ is the mean free path), $\Omega_i[\ \cdot\ ]$ is the collision operator representing the increase or decrease of the velocity distribution function f_i due to the collision of fictitious particles, $\boldsymbol{f} = (f_1, f_2, \cdots, f_N)^{\mathrm{T}}$, and N ($= 9$ or 15) is the number of particle velocities. The left-hand and right-hand sides of the above equation represent the free streaming and collisions of the fictitious particles, respectively. In the isothermal model, the collision term must satisfy the following equations from the law of conservation of mass and momentum:

$$\sum_{i=1}^{N} \Omega_i = 0, \qquad (1.4)$$

$$\sum_{i=1}^{N} \boldsymbol{c}_i \Omega_i = \boldsymbol{0}. \qquad (1.5)$$

Next, the physical space is divided into a square lattice (in the D2Q9 model) or a cubic lattice (in the D3Q15 model) with lattice spacing Δx, and the time step is set as $\Delta t = \mathrm{Sh}\Delta x$. In one time step, a fictitious particle moves to the next lattice point.[5] Approximating the left-hand side of Eq. (1.3) by the first-order difference in the characteristic direction, we obtain the following equation (Fig. 1.2):

$$f_i(\boldsymbol{x} + \boldsymbol{c}_i\Delta x, t + \Delta t) - f_i(\boldsymbol{x}, t) = \Omega_i[\boldsymbol{f}(\boldsymbol{x}, t)], \qquad i = 1, 2, \cdots, N, \qquad (1.6)$$

where ε is set to Δx. In addition, as $\mathrm{Sh} = O(\Delta x)$ on the diffusive time scale (see Appendix A), we have $\Delta t = O((\Delta x)^2)$.

Equation (1.6), called the lattice Boltzmann equation, is the basic equation of the LBM. Using the lattice Boltzmann equation, we can explicitly

---

[4]The circumflex $\hat{\ }$ denotes a dimensional variable as described in Appendix A.

[5]Since $\hat{c}\Delta\hat{t} = \Delta\hat{x}$ in the dimensional form is expressed as $\hat{c}\Delta t(\hat{L}/\hat{U}) = \Delta x\hat{L}$ in dimensionless form, we clearly have $\Delta t = (\hat{U}/\hat{c})\Delta x = \mathrm{Sh}\Delta x$.

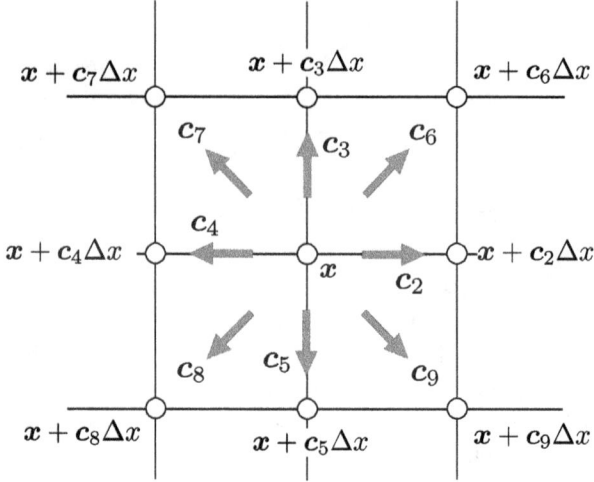

Fig. 1.2　Computational lattice and particle velocities in the D2Q9 model.

compute the time evolution of the velocity distribution function f_i. As shown in Fig. 1.2, the velocity distribution function of a lattice point at a new time is determined only by the velocity distribution function of the neighboring lattice point from which the fictitious particle departed. Namely, the velocity distribution function is perfectly advected. Also, from Eqs. (1.4) and (1.5), we confirm that mass and momentum are conserved at the lattice points before and after collision. Therefore, the LBM excellently conserves both mass and momentum. However, as the diffusive time scale is $\Delta t = O((\Delta x)^2)$, many (several hundred) time steps are needed to examine the slow changes in physical quantities.

The collision term on the right-hand side takes a complicated form. For example, for a stationary particle ($i = 1$) in the D2Q9 model limited to two-body collisions, this term becomes [56]

$$\Omega_1 = R[(f_2 f_3 - f_1 f_6) + (f_3 f_4 - f_1 f_7) + (f_4 f_5 - f_1 f_8) + (f_2 f_5 - f_1 f_9)], \quad (1.7)$$

where R is a constant proportional to the relative velocity difference between a moving particle and the stationary particle with which it collides.

The macroscopic fluid density ρ and flow velocity u are respectively determined from the velocity distribution function f_i as follows:

$$\rho = \sum_{i=1}^{N} f_i, \quad (1.8)$$

$$u = \frac{1}{\rho} \sum_{i=1}^{N} c_i f_i. \tag{1.9}$$

Note that in the LBM of an incompressible viscous fluid, $|u| = O(\Delta x)$ as explained later (see Sec. 1.5).

Meanwhile, the pressure p is defined as

$$p = \frac{2}{D} \rho e, \tag{1.10}$$

where D is the number of spatial dimensions ($D = 2$ and 3 in the two- and three-dimensional cases, respectively), and the internal energy e is calculated as

$$e = \frac{1}{2\rho} \sum_{i=1}^{N} |c_i - u|^2 f_i. \tag{1.11}$$

Note that because the model is isothermal, the internal-energy calculation can be omitted and we can derive $e = D/6$ and $p = \rho/3$ as described in Sec. 1.5. Therefore, the speed of sound in the LBM becomes $c_s = \sqrt{dp/d\rho} = \sqrt{1/3}$ (as also confirmed in Sec. 1.9).

In summary, the LBM computes the time evolution of the velocity distribution function f_i on the fast (acoustic) time scale, during which a fictitious particle moves to the next lattice point. Meanwhile, the density ρ (or pressure p) and flow velocity u are tracked by computing the moments of f_i on the slow (diffusive) time scale.

1.3 Formulation of the Collision Term

Handling of collision terms is a key issue in the lattice Boltzmann equation as well as in the general Boltzmann equation. In the early stage of LBM development, the collision term was described by a collision scattering matrix model [49, 96]. Later improvements of the scattering matrix model include the single-relaxation-time (SRT) model [14, 109] based on the Bhatnagar–Gross–Krook (BGK) model [7] and the multiple-relaxation-time (MRT) model [23, 88]. This book focuses on the collision scattering matrix model and the BGK (SRT) model. The MRT model is described in Appendix C.

(1) Collision scattering matrix model

In the collision scattering matrix model, the collision term on the right-hand side of Eq. (1.6) is expressed as follows:

$$\Omega_i = \sum_{j=1}^{N} A_{ij}[f_j(\boldsymbol{x},t) - f_j^{\text{eq}}(\boldsymbol{x},t)], \qquad (1.12)$$

where A_{ij} is the collision scattering matrix, and f_j^{eq} is the local equilibrium distribution function. The above expression is obtained by expanding the collision term on the right-hand side of Eq. (1.6) around the local equilibrium distribution and retaining only the first term, i.e., $A_{ij} \equiv \partial \Omega_i / \partial f_j(\boldsymbol{f}^{\text{eq}})$.[6] The matrix A_{ij} is determined from the transition probability between a fictitious particle with velocity $\boldsymbol{c}_i$ and a fictitious particle with velocity $\boldsymbol{c}_j$ due to collision. Note that A_{ij} is a real symmetric matrix. Therefore, it has N real eigenvalues (including duplications) and N linearly independent eigenvectors (orthogonal vectors can be constructed). The eigenvalues and eigenvectors of A_{ij} correspond to the physical quantities conserved by collision (eigenvalues = 0) or the physical quantities that relax to equilibrium (eigenvalues $\neq$ 0). In the isothermal model, the conserved physical quantities are mass and momentum and the physical quantities that relax to the equilibrium state are stress tensors. The other quantities that relax to the equilibrium state generally include non-physical quantities. Recall that the left-hand side of Eq. (1.6) describes perfect advection of the velocity distribution function (see Fig. 1.2). Therefore, the numerical scheme given by Eqs. (1.6), (1.8), and (1.9) conserves the mass ρ and momentum $\rho\boldsymbol{u}$, clarifying that the mass and momentum are excellently conserved in the LBM.

In the collision term model, the eigenvalue corresponding to the stress tensor of the matrix A_{ij} is inversely proportional to the fluid viscosity, but the viscosity coefficient of the fluid cannot be reduced, i.e., this model is limited to flows with moderate Reynolds numbers (~ 100). Flows with high Reynolds number cannot be computed because the number of velocities of the fictitious particle is limited to N and the change rate of a physical quantity that relaxes to the equilibrium state through collisions is small. For this reason, researchers have attempted to modify the eigenvalues that determine viscosity.

---

[6]Note that the collision term in Eq. (1.12) is nonlinear because f_j^{eq} is generally nonlinear with respect to f_j.

The simplest attempt is the BGK (SRT) model [14, 109], which constructs A_{ij} as a diagonal matrix with equal eigenvalues.

(2) The BGK (SRT) model

Assuming $A_{ij} = -\frac{1}{\tau}\delta_{ij}$ in Eq. (1.12), where δ_{ij} is the Kronecker delta, we obtain the following diagonal matrix with equal eigenvalues:

$$\Omega_i = -\frac{1}{\tau}[f_i(\boldsymbol{x},t) - f_i^{\text{eq}}(\boldsymbol{x},t)], \tag{1.13}$$

where $\tau = O(1)$ is called a relaxation time. The above equation corresponds to the BGK collision model [7] in the general Boltzmann equation. Moreover, as the relaxation time is a single value (τ), this model is also called the SRT model.

From Eqs. (1.6) and (1.13), the basic equation of the LBM with the BGK model is given by

$$f_i(\boldsymbol{x} + \boldsymbol{c}_i\Delta x, t + \Delta t) = f_i(\boldsymbol{x},t) - \frac{1}{\tau}[f_i(\boldsymbol{x},t) - f_i^{\text{eq}}(\boldsymbol{x},t)]. \tag{1.14}$$

Although the above equation is a simple time evolution equation, it is nonlinear because f_i^{eq} is nonlinear with respect to f_i (Sec. 1.4).[7] In this book, the LBM is mainly formulated using the BGK model, and is called the BGK-LBM or SRT-LBM.

The numerical stability of the SRT model has been improved by two ideas: the regularized LBM (RLBM) [90] that corrects the non-equilibrium part of the velocity distribution function f_i, and the entropic LBM (ELBM) [4] that accounts for the entropy-increasing law. These improvements are not further discussed in this book.

1.4 Local Equilibrium Distribution Function

The local equilibrium distribution function f_i^{eq} on the right-hand side of Eq. (1.14) must satisfy the following constraints:

$$\sum_{i=1}^{N} f_i^{\text{eq}} = \rho, \tag{1.15}$$

$$\frac{1}{\rho}\sum_{i=1}^{N} \boldsymbol{c}_i f_i^{\text{eq}} = \boldsymbol{u}. \tag{1.16}$$

---

[7]The basic equation of the LBM is sometimes mistaken for a linear equation.

One possible solution is [109]

$$f_i^{\text{eq}} = E_i \rho \left[1 + 3c_i \cdot u + \frac{9}{2}(c_i \cdot u)^2 - \frac{3}{2} u \cdot u \right], \tag{1.17}$$

where ρ and u at the right-hand side are obtained from f_i using Eqs. (1.8) and (1.9), respectively. The coefficients E_i are given by

$$E_i = \begin{cases} \frac{4}{9}, & i = 1, \\ \frac{1}{9}, & i = 2, 3, 4, 5, \\ \frac{1}{36}, & i = 6, 7, 8, 9, \end{cases} \tag{1.18}$$

in the two-dimensional nine-velocity model and

$$E_i = \begin{cases} \frac{2}{9}, & i = 1, \\ \frac{1}{9}, & i = 2, 3, \cdots, 7, \\ \frac{1}{72}, & i = 8, 9, \cdots, 15, \end{cases} \tag{1.19}$$

in the three-dimensional fifteen-velocity model.

The expression in parentheses on the right-hand side of Eq. (1.17) is obtained by expanding the Maxwell distribution [126], which is a local equilibrium solution of general gas molecule kinetics, on the assumption that the flow speed $|u|$ is small and by retaining the terms up to $|u|^2$. However, it is unclear whether the Maxwell distribution holds for the lattice gas model. Therefore, this expansion is only an assumption. The coefficients E_i are determined such that Eq. (1.15), Eq. (1.16), and the stress tensors of the Navier–Stokes equations are all satisfied [1, 45]. In the isothermal model, the coefficients E_i are constant. Equation (1.17) is generally used, but neither the coefficients E_i nor the coefficients in parentheses on the right-hand side of Eq. (1.17) are uniquely determined.

Equations (1.14), (1.17), (1.8), and (1.9) constitute the numerical scheme of the LBM. In a linear stability analysis, this numerical scheme is confirmed to be stable when $\tau > 1/2$ [128]. This corresponds to the positive kinematic viscosity coefficient ν, as can be seen from Eq. (1.49) in the next section. However, as Eq. (1.14) is nonlinear, the LBM becomes unstable due to the nonlinear effect near $\tau = 1/2$. The numerical instability of the LBM is shown in a numerical example in Sec. 2.7.

1.5   Derivation of the Fluid Dynamic Equations

The macroscopic variables u and p $(= \rho/3)$ computed using Eqs. (1.14), (1.17), (1.8), and (1.9) approximately satisfy the continuity equation and the Navier–Stokes equations for incompressible fluids. This agreement is usually confirmed by a Chapman–Enskog expansion [15], but is here confirmed by Sone's asymptotic analysis (S-expansion) [125], which is more easily understood. Whereas the Chapman–Enskog expansion considers both the diffusive and acoustic time scales, the S-expansion considers only the changes on the diffusive time scale as shown below [58,78]. The S-expansion expands both f_i and the macroscopic variables as power series of Δx, whereas the Chapman–Enskog expansion expands only f_i as a power series of Δx. Therefore, the orders of the macroscopic variables are more easily compared in the S-expansion than in the Chapman–Enskog expansion. The following derivation is demonstrated in the two-dimensional nine-velocity model, but can also be confirmed in the three-dimensional fifteen-velocity model.

In general, the relation[8] Ma ~ Kn × Re holds among the Mach number Ma = $\hat{U}/\hat{c}_s$ (where $\hat{c}_s$ is the speed of sound), the Knudsen number Kn = $\hat{\ell}/\hat{L}$ (where $\hat{\ell}$ is the mean free path), and the Reynolds number Re = $\hat{\rho}\hat{U}\hat{L}/\hat{\mu}$ (where $\hat{\mu}$ is the viscosity coefficient). Therefore, when the Reynolds number is finite and the Knudsen number is small (corresponding to the flow of an incompressible viscous fluid), the Mach number becomes as small as the Knudsen number. Given that the deviation from the stationary equilibrium state is of the same order as the Knudsen number $O(\Delta x)$, we can expand f_i as follows:

$$f_i = E_i[1 + (\Delta x)f_i^{(1)} + (\Delta x)^2 f_i^{(2)} + (\Delta x)^3 f_i^{(3)} + (\Delta x)^4 f_i^{(4)} + \cdots]. \quad (1.20)$$

Corresponding to the above expansion, the macroscopic variables are then expanded as follows:[9]

$$\rho = 1 + (\Delta x)\rho^{(1)} + (\Delta x)^2\rho^{(2)} + (\Delta x)^3\rho^{(3)} + \cdots, \quad (1.21)$$

$$\boldsymbol{u} = (\Delta x)\boldsymbol{u}^{(1)} + (\Delta x)^2\boldsymbol{u}^{(2)} + (\Delta x)^3\boldsymbol{u}^{(3)} + \cdots, \quad (1.22)$$

$$e = \frac{1}{3} + (\Delta x)e^{(1)} + (\Delta x)^2 e^{(2)} + (\Delta x)^3 e^{(3)} + \cdots, \quad (1.23)$$

$$p = \frac{1}{3} + (\Delta x)p^{(1)} + (\Delta x)^2 p^{(2)} + (\Delta x)^3 p^{(3)} + \cdots. \quad (1.24)$$

Note that $|\boldsymbol{u}| = O(\Delta x)$ in the above equation. This is an important requirement of the LBM; that is, the flow velocity must be smaller than the velocity of the fictitious particle.

---

[8]It is called the von Kármán relation. From the relation $\hat{\mu} \sim \hat{\rho}\hat{c}_s\hat{\ell}$, we find that Re = $\hat{\rho}\hat{U}\hat{L}/\hat{\mu} \sim (\hat{\rho}\hat{U}\hat{L})/(\hat{\rho}\hat{c}_s\hat{\ell}) \sim$ Ma/Kn [126].

[9]In three-dimensional cases, note that $e = \frac{1}{2} + (\Delta x)e^{(1)} + (\Delta x)^2 e^{(2)} + (\Delta x)^3 e^{(3)} + \cdots$.

Next, performing a Taylor expansion of Eq. (1.14) around $(\boldsymbol{x}, t)$ up to the order of $(\Delta x)^3$, we have

$$(\Delta x)\left(\mathrm{Sh}\frac{\partial}{\partial t} + c_{i\alpha}\frac{\partial}{\partial x_\alpha}\right)f_i + \frac{1}{2}(\Delta x)^2\left(\mathrm{Sh}\frac{\partial}{\partial t} + c_{i\alpha}\frac{\partial}{\partial x_\alpha}\right)^2 f_i$$

$$+\frac{1}{6}(\Delta x)^3\left(c_{i\alpha}\frac{\partial}{\partial x_\alpha}\right)^3 f_i + O((\Delta x)^5) = -\frac{1}{\tau}(f_i - f_i^{\mathrm{eq}}), \qquad (1.25)$$

where the subscript $\alpha = x, y$ represents Cartesian coordinates in the summation convention.[10]

Here, we consider a moderately varying solution, i.e., $\partial f_i/\partial t = O(f_i)$ and $\partial f_i/\partial x_\alpha = O(f_i)$. Substituting Eqs. (1.20)–(1.24) into Eq. (1.25) and collecting terms with the same order of Δx (note that $\mathrm{Sh} = O(\Delta x)$ on the diffusive time scale), we get

$$f_i^{(1)} = \rho^{(1)} + 3c_{i\alpha}u_\alpha^{(1)}, \qquad (1.26)$$

$$f_i^{(2)} = \rho^{(2)} + 3\rho^{(1)}c_{i\alpha}u_\alpha^{(1)} + 3c_{i\alpha}u_\alpha^{(2)}$$

$$+\frac{9}{2}c_{i\alpha}c_{i\beta}u_\alpha^{(1)}u_\beta^{(1)} - \frac{3}{2}u_\alpha^{(1)}u_\alpha^{(1)} - \tau c_{i\alpha}\frac{\partial f_i^{(1)}}{\partial x_\alpha}, \qquad (1.27)$$

$$f_i^{(3)} = \rho^{(3)} + 3\rho^{(2)}c_{i\alpha}u_\alpha^{(1)} + 3\rho^{(1)}c_{i\alpha}u_\alpha^{(2)} + 3c_{i\alpha}u_\alpha^{(3)}$$

$$+\frac{9}{2}\rho^{(1)}c_{i\alpha}c_{i\beta}u_\alpha^{(1)}u_\beta^{(1)} + 9c_{i\alpha}c_{i\beta}u_\alpha^{(1)}u_\beta^{(2)}$$

$$-\frac{3}{2}\rho^{(1)}u_\alpha^{(1)}u_\alpha^{(1)} - 3u_\alpha^{(1)}u_\alpha^{(2)}$$

$$-\tau c_{i\alpha}\frac{\partial f_i^{(2)}}{\partial x_\alpha} - \tau\frac{\mathrm{Sh}}{\Delta x}\frac{\partial f_i^{(1)}}{\partial t} - \frac{1}{2}\tau c_{i\alpha}c_{i\beta}\frac{\partial^2 f_i^{(1)}}{\partial x_\alpha\partial x_\beta}, \qquad (1.28)$$

---

[10]We apply the same rules to the subscripts β, γ, and δ. In the three-dimensional fifteen-velocity model, $\alpha, \beta, \gamma, \delta = x, y, z$.

$$f_i^{(4)} = \rho^{(4)} + 3\rho^{(3)} c_{i\alpha} u_\alpha^{(1)} + 3\rho^{(2)} c_{i\alpha} u_\alpha^{(2)} + 3\rho^{(1)} c_{i\alpha} u_\alpha^{(3)} + 3 c_{i\alpha} u_\alpha^{(4)}$$

$$+ \frac{9}{2} \rho^{(2)} c_{i\alpha} c_{i\beta} u_\alpha^{(1)} u_\beta^{(1)} + 9\rho^{(1)} c_{i\alpha} c_{i\beta} u_\alpha^{(1)} u_\beta^{(2)}$$

$$+ 9 c_{i\alpha} c_{i\beta} u_\alpha^{(1)} u_\beta^{(3)} + \frac{9}{2} c_{i\alpha} c_{i\beta} u_\alpha^{(2)} u_\beta^{(2)}$$

$$- \frac{3}{2} \rho^{(2)} u_\alpha^{(1)} u_\alpha^{(1)} - 3\rho^{(1)} u_\alpha^{(1)} u_\alpha^{(2)} - \frac{3}{2} u_\alpha^{(2)} u_\alpha^{(2)} - 3 u_\alpha^{(1)} u_\alpha^{(3)}$$

$$- \tau c_{i\alpha} \frac{\partial f_i^{(3)}}{\partial x_\alpha} - \tau \frac{Sh}{\Delta x} \frac{\partial f_i^{(2)}}{\partial t} - \frac{1}{2} \tau c_{i\alpha} c_{i\beta} \frac{\partial^2 f_i^{(2)}}{\partial x_\alpha \partial x_\beta}$$

$$- \tau \frac{Sh}{\Delta x} c_{i\alpha} \frac{\partial^2 f_i^{(1)}}{\partial t \partial x_\alpha} - \frac{1}{6} \tau c_{i\alpha} c_{i\beta} c_{i\gamma} \frac{\partial^3 f_i^{(1)}}{\partial x_\alpha \partial x_\beta \partial x_\gamma}. \tag{1.29}$$

The above equations are systems of linear equations for the unknowns $f_i^{(m)}$ $(m = 1, 2, \cdots)$. They can be summarized as follows:

$$f_i^{(1)} - \sum_{j=1}^{9} E_j f_j^{(1)} - 3 c_{i\alpha} \sum_{j=1}^{9} E_j c_{j\alpha} f_j^{(1)} = 0, \tag{1.30}$$

$$f_i^{(m)} - \sum_{j=1}^{9} E_j f_j^{(m)} - 3 c_{i\alpha} \sum_{j=1}^{9} E_j c_{j\alpha} f_j^{(m)} = Ih_i^{(m)} \quad (m \geq 2), \tag{1.31}$$

where the inhomogeneous terms $Ih_i^{(m)}$ $(m \geq 2)$ consist of the second and subsequent rows on the right-hand side of Eqs. (1.27), (1.28), and (1.29).

Now, Eq. (1.30) is a homogeneous system of linear equations, and its solution is given by Eq. (1.26). Substituting Eq. (1.26) into Eq. (1.11), we then get

$$e^{(1)} = 0. \tag{1.32}$$

Conversely, Eq. (1.31) is an inhomogeneous system of linear equations with the same constant coefficients regardless of m. In the following, we will examine the solvability condition of Eq. (1.31) based on this constant coefficient matrix, which is denoted by P and explicitly written as follows:

$$P = \frac{1}{36} \begin{bmatrix} 20 & -4 & -4 & -4 & -4 & -1 & -1 & -1 & -1 \\ -16 & 20 & -4 & 8 & -4 & -4 & 2 & 2 & -4 \\ -16 & -4 & 20 & -4 & 8 & -4 & -4 & 2 & 2 \\ -16 & 8 & -4 & 20 & -4 & 2 & -4 & -4 & 2 \\ -16 & -4 & 8 & -4 & 20 & 2 & 2 & -4 & -4 \\ -16 & -16 & -16 & 8 & 8 & 29 & -1 & 5 & -1 \\ -16 & 8 & -16 & -16 & 8 & -1 & 29 & -1 & 5 \\ -16 & 8 & 8 & -16 & -16 & 5 & -1 & 29 & -1 \\ -16 & -16 & 8 & 8 & -16 & -1 & 5 & -1 & 29 \end{bmatrix}. \tag{1.33}$$

It is easily verified that rank(P) = 6 and its homogeneous equation has three nontrivial solutions: $(1,1,1,1,1,1,1,1,1)^{\mathrm{T}}$, $(0,1,0,-1,0,1,-1,-1,1)^{\mathrm{T}}$, and $(0,0,1,0,-1,1,1,-1,-1)^{\mathrm{T}}$. Note that the latter two solutions correspond to $c_{i\alpha}$. It is also easily seen that after multiplying each row of the matrix P by E_i, the resulting matrix is symmetric. Thus, from the fundamental theorem [11] [129] of linear algebra, we obtain the solvability conditions of Eq. (1.31) as follows:

$$\sum_{i=1}^{9} E_i I h_i^{(m)} = 0 \quad (m \geq 2), \tag{1.34}$$

$$\sum_{i=1}^{9} E_i c_{i\alpha} I h_i^{(m)} = 0 \quad (m \geq 2). \tag{1.35}$$

Now, let us check the above solvability conditions for m = 2 and higher. (At this time, the formulae in Appendix D are useful.)

First, under the solvability condition for m = 2 we have

$$\frac{\partial u_{\alpha}^{(1)}}{\partial x_{\alpha}} = 0, \tag{1.36}$$

$$\frac{\partial \rho^{(1)}}{\partial x_{\alpha}} = 0. \tag{1.37}$$

Substituting Eq. (1.27) into Eq. (1.11), we then get

$$e^{(2)} = 0. \tag{1.38}$$

Next, under the solvability condition for m = 3 we have

---

[11] $P\boldsymbol{x} = \boldsymbol{b}$ has a solution if and only if $\boldsymbol{b}$ is orthogonal to all solutions $\boldsymbol{y}$ of $P^{\mathrm{T}}\boldsymbol{y} = \boldsymbol{0}$.

$$\frac{Sh}{\Delta x}\frac{\partial \rho^{(1)}}{\partial t} + \frac{\partial u_\alpha^{(2)}}{\partial x_\alpha} = 0, \tag{1.39}$$

$$\frac{Sh}{\Delta x}\frac{\partial u_\alpha^{(1)}}{\partial t} + u_\beta^{(1)}\frac{\partial u_\alpha^{(1)}}{\partial x_\beta} = -\frac{1}{3}\frac{\partial \rho^{(2)}}{\partial x_\alpha}$$

$$+\frac{1}{3}\left(\tau - \frac{1}{2}\right)\left(\frac{\partial^2 u_\alpha^{(1)}}{\partial x_\beta^2} + 2\frac{\partial^2 u_\beta^{(1)}}{\partial x_\alpha \partial x_\beta}\right). \tag{1.40}$$

Substituting Eq. (1.28) into Eq. (1.11), we then get

$$e^{(3)} = -\frac{1}{3}\tau\frac{\partial u_\alpha^{(2)}}{\partial x_\alpha}. \tag{1.41}$$

Finally, under the solvability condition for $m = 4$ we have

$$\frac{Sh}{\Delta x}\frac{\partial \rho^{(2)}}{\partial t} + \frac{\partial}{\partial x_\alpha}(\rho^{(1)}u_\alpha^{(2)} + \rho^{(2)}u_\alpha^{(1)} + u_\alpha^{(3)}) = 0, \tag{1.42}$$

$$\frac{Sh}{\Delta x}\frac{\partial}{\partial t}(\rho^{(1)}u_\alpha^{(1)} + u_\alpha^{(2)}) + \frac{\partial}{\partial x_\beta}(\rho^{(1)}u_\alpha^{(1)}u_\beta^{(1)} + u_\alpha^{(1)}u_\beta^{(2)} + u_\beta^{(1)}u_\alpha^{(2)})$$

$$= -\frac{1}{3}\frac{\partial \rho^{(3)}}{\partial x_\alpha}$$

$$+\frac{1}{3}\left(\tau - \frac{1}{2}\right)\left[\frac{\partial^2}{\partial x_\beta^2}(\rho^{(1)}u_\alpha^{(1)} + u_\alpha^{(2)}) + 2\frac{\partial^2}{\partial x_\alpha \partial x_\beta}(\rho^{(1)}u_\beta^{(1)} + u_\beta^{(2)})\right]. \tag{1.43}$$

Now, from Eq. (1.37), we find that $\rho^{(1)}$ is constant in space. Thus, if the boundary condition for $\rho^{(1)}$ is time-independent, then $\rho^{(1)}$ is a constant independent of time and space. Incorporating this constant into the reference density ρ_0, we can therefore set $\rho^{(1)} = 0$ without loss of generality. From Eqs. (1.36) and (1.40), the governing equations for $u_\alpha^{(1)}$ and $p^{(2)}(= \rho^{(2)}/3)$ are respectively obtained as

$$\frac{\partial u_\alpha^{(1)}}{\partial x_\alpha} = 0, \tag{1.44}$$

$$\frac{Sh}{\Delta x}\frac{\partial u_\alpha^{(1)}}{\partial t} + u_\beta^{(1)}\frac{\partial u_\alpha^{(1)}}{\partial x_\beta} = -\frac{\partial p^{(2)}}{\partial x_\alpha} + \frac{1}{3}\left(\tau - \frac{1}{2}\right)\frac{\partial^2 u_\alpha^{(1)}}{\partial x_\beta^2}. \tag{1.45}$$

The above equations clarify that $u_\alpha^{(1)}$ and $p^{(2)}$ satisfy the continuity equation and the Navier–Stokes equations for incompressible viscous fluids.[12]

---

[12]In the Chapman–Enskog expansion, deriving the equations of compressible viscous fluids and then assuming a low Mach number flow, we obtain the equations of incompressible viscous fluids [15].

Moreover, from Eqs. (1.36), (1.39), and (1.43) we get the following governing equations for $u_\alpha^{(2)}$ and $p^{(3)}(= \rho^{(3)}/3)$:

$$\frac{\partial u_\alpha^{(2)}}{\partial x_\alpha} = 0, \tag{1.46}$$

$$\frac{\text{Sh}}{\Delta x}\frac{\partial u_\alpha^{(2)}}{\partial t} + u_\beta^{(1)}\frac{\partial u_\alpha^{(2)}}{\partial x_\beta} + u_\beta^{(2)}\frac{\partial u_\alpha^{(1)}}{\partial x_\beta} = -\frac{\partial p^{(3)}}{\partial x_\alpha} + \frac{1}{3}\left(\tau - \frac{1}{2}\right)\frac{\partial^2 u_\alpha^{(2)}}{\partial x_\beta^2}. \tag{1.47}$$

Equations (1.46) and (1.47) are homogeneous linear simultaneous equations for $u_\alpha^{(2)}$ and $p^{(3)}$. If both the initial and boundary conditions of $u_\alpha^{(2)}$ are 0 (that is, the initial and boundary conditions of u_α are independent of Δx and are imposed only on the first term $u_\alpha^{(1)}$), we easily see that $u_\alpha^{(2)} = p^{(3)} = 0$ in all regions. Therefore, $u_\alpha = (\Delta x)u_\alpha^{(1)} + O((\Delta x)^3)$ and $p = 1/3 + (\Delta x)^2 p^{(2)} + O((\Delta x)^4)$ satisfy the continuity equation (1.44) and the Navier–Stokes equations (1.45) for incompressible viscous fluids with relative errors $O((\Delta x)^2)$. It is also found that[13]

$$e = \frac{1}{3} + O((\Delta x)^4), \tag{1.48}$$

indicating that the flow is isothermal up to the order of $(\Delta x)^3$. Multiplying both sides of Eq. (1.45) by $(\Delta x)^2$ restores the original magnitudes of the macroscopic variables. The kinematic viscosity coefficient ν is then given by

$$\nu = \frac{1}{3}\left(\tau - \frac{1}{2}\right)\Delta x. \tag{1.49}$$

Thus, to reduce ν in the flow calculations at high Reynolds numbers, we should either reduce Δx or bring τ closer to $1/2$.

The stress tensor $\sigma_{\alpha\beta}$ in the fluid can be obtained from the velocity distribution functions f_i [62] as follows:

$$\sigma_{\alpha\beta} = -\frac{1}{2\tau}p\delta_{\alpha\beta} - \frac{\tau - \frac{1}{2}}{\tau}\sum_{i=1}^{N}f_i(c_{i\alpha} - u_\alpha)(c_{i\beta} - u_\beta). \tag{1.50}$$

From the asymptotic analysis (S-expansion), we obtain $\sigma_{\alpha\beta} = (1/3)\delta_{\alpha\beta} + (\Delta x)^2\sigma_{\alpha\beta}^{(2)} + O((\Delta x)^4)$, indicating that $\sigma_{\alpha\beta}$ can also be calculated with relative errors $O((\Delta x)^2)$.[14]

---

[13]In three dimensions, $e = \frac{1}{2} + O((\Delta x)^4)$.

[14]The constant term $(1/3)\delta_{\alpha\beta}$ does not contribute to the velocity and pressure fields.

To summarize, the velocity u and pressure p $(= \rho/3)$ of incompressible viscous fluids in an isothermal field can be calculated using Eqs. (1.14), (1.17), (1.8), and (1.9) with relative errors $O((\Delta x)^2)$.[15] Thus, the LBM can be considered to be a numerical method of the second-order spatial accuracy for incompressible viscous fluids.[16] However, the above analysis assumes that the velocity distribution function and the macroscopic variables are smooth everywhere, including in the initial state and on the boundary. It should be noted that if the initial and boundary conditions of the velocity distribution function are poorly applied, discontinuous changes called the initial layer and the Knudsen layer generally occur, which decrease the numerical accuracy [79].

As mentioned above, the LBM can be a numerical method of the second-order spatial accuracy for incompressible fluids, but the three-dimensional fifteen-velocity model has 15 unknown functions, whereas the usual three-dimensional incompressible Navier–Stokes equations include four unknowns of the velocity u and pressure p. In addition, to follow phenomena on the diffusive time scale, a large number of time steps is required. However, because LBM does not need to solve the Poisson equation of pressure, its computational time is not considerably increased from that of conventional numerical schemes for the incompressible Navier–Stokes equations. The LBM additionally provides a simple numerical scheme, excellent conservation of mass and momentum, and suitability for parallel computing (see, e.g. [119]). In recent years, massive parallel computing of LBMs has been performed on GPUs [156].

As a numerical scheme with weak compressibility, the LBM can be regarded as an artificial compressibility method. The relationship between the LBM and artificial compressibility methods is detailed elsewhere (see, e.g. [48, 102, 103]).

1.6   LBM with an External Force Term

The LBM for the Navier–Stokes equations with an external force term can be written by [43]

---

[15]In Eq. (1.42), the compressibility effect $\partial \rho^{(2)}/\partial t \neq 0$ appears in the macroscopic variable $u_\alpha^{(3)}$. That is, the compressibility effect is embodied in the relative error $O((\Delta x)^2)$.
[16]Although the temporal accuracy is $O(\Delta t)$, a relative error $O((\Delta x)^2)$ is guaranteed because $\Delta t = \mathrm{Sh}\Delta x = O((\Delta x)^2)$.

$$f_i(\boldsymbol{x} + \boldsymbol{c}_i \Delta x, t + \Delta t) = f_i(\boldsymbol{x}, t) - \frac{1}{\tau}[f_i(\boldsymbol{x}, t) - f_i^{\text{eq}}(\boldsymbol{x}, t)]$$

$$+ 3\Delta x E_i \boldsymbol{c}_i \cdot \boldsymbol{G}(\boldsymbol{x}, t). \tag{1.51}$$

In the above equation, $\boldsymbol{G}(\boldsymbol{x}, t)$ ($|\boldsymbol{G}| = O((\Delta x)^2)$) is the external force acting per unit mass.[17] Now, expanding $\boldsymbol{G}$ as

$$\boldsymbol{G} = (\Delta x)^2 \boldsymbol{G}^{(2)} + (\Delta x)^3 \boldsymbol{G}^{(3)} + \cdots, \tag{1.52}$$

and performing the analysis described in the previous section, the governing equations of $u_\alpha^{(1)}$ and $p^{(2)}$ ($= \rho^{(2)}/3$) are derived as follows:

$$\frac{\partial u_\alpha^{(1)}}{\partial x_\alpha} = 0, \tag{1.53}$$

$$\frac{\text{Sh}}{\Delta x} \frac{\partial u_\alpha^{(1)}}{\partial t} + u_\beta^{(1)} \frac{\partial u_\alpha^{(1)}}{\partial x_\beta} = -\frac{\partial p^{(2)}}{\partial x_\alpha} + \frac{1}{3}\left(\tau - \frac{1}{2}\right)\frac{\partial^2 u_\alpha^{(1)}}{\partial x_\beta^2} + G_\alpha^{(2)}, \tag{1.54}$$

which are the continuity equation and the Navier–Stokes equations with the external force term, respectively. Next, the governing equations of $u_\alpha^{(2)}$ and $p^{(3)}$ ($= \rho^{(3)}/3$) are given by

$$\frac{\partial u_\alpha^{(2)}}{\partial x_\alpha} = 0, \tag{1.55}$$

$$\frac{\text{Sh}}{\Delta x}\frac{\partial u_\alpha^{(2)}}{\partial t} + u_\beta^{(1)}\frac{\partial u_\alpha^{(2)}}{\partial x_\beta} + u_\beta^{(2)}\frac{\partial u_\alpha^{(1)}}{\partial x_\beta} = -\frac{\partial p^{(3)}}{\partial x_\alpha} + \frac{1}{3}\left(\tau - \frac{1}{2}\right)\frac{\partial^2 u_\alpha^{(2)}}{\partial x_\beta^2}$$

$$+ G_\alpha^{(3)}. \tag{1.56}$$

Therefore, if $G_\alpha^{(3)} = 0$, meaning that G_α does not depend on Δx and is determined only by the first term $G_\alpha^{(2)}$, or if $G_\alpha = (\Delta x)^2 G_\alpha^{(2)} + O((\Delta x)^4)$, then (as similarly argued in the previous section) the velocity $\boldsymbol{u}$ and pressure p ($= \rho/3$) can be calculated with relative errors $O((\Delta x)^2)$.

The additional term $(1/2)(\partial G_\alpha^{(2)}/\partial x_\alpha)$ appears on the left-hand side of Eq. (1.42), but does not affect the above conclusion because its effect is involved in relative error $O((\Delta x)^2)$. It has been pointed out that in the Chapman–Enskog expansion, the spatial and temporal derivatives of the external force term $\boldsymbol{G}(\boldsymbol{x}, t)$ in Eq. (1.51) reduce the numerical accuracy

---

[17]For example, to ensure a finite Froude number $\text{Fr} = U/\sqrt{gL}$, the external force (gravitational acceleration g) should be $O(U^2) = O((\Delta x)^2)$.

of the scheme [42]. However, according to the above asymptotic analysis (S-expansion), relative errors $O((\Delta x)^2)$ are guaranteed when $G(x,t)$ is smooth [141]. The Chapman–Enskog expansion does not expand the macroscopic variables as a power series of Δx, so the magnitude of $G(x,t)$ is not known. In contrast, the orders of the macroscopic variables can be clarified in the asymptotic analysis (S-expansion).

1.7 Incompressible Local Equilibrium Distribution Function

As shown in the above explanation, $p = \rho/3$ in the LBM for isothermal fluids, implying a one-to-one correspondence between density and pressure. Therefore, the pressure can replace the density as a macroscopic variable. Accordingly, we often use the local equilibrium distribution function with a slight modification as follows [44, 167]:

$$f_i(x + c_i\Delta x, t + \Delta t) = f_i(x,t) - \frac{1}{\tau}[f_i(x,t) - f_i^{\text{eq,in}}(x,t)], \qquad (1.57)$$

where

$$f_i^{\text{eq,in}} = E_i\left[3p + 3c_i \cdot u + \frac{9}{2}(c_i \cdot u)^2 - \frac{3}{2}u \cdot u\right], \qquad (1.58)$$

$$p = \frac{1}{3}\sum_{i=1}^{N} f_i, \qquad (1.59)$$

$$u = \sum_{i=1}^{N} c_i f_i. \qquad (1.60)$$

Applying the asymptotic analysis (S-expansion) in Sec. 1.5 to Eqs. (1.57)–(1.60), we find that u and p satisfy the continuity equation and the Navier–Stokes equations with relative errors $O((\Delta x)^2)$. Although $f_i^{\text{eq,in}}$ is also called the incompressible local equilibrium distribution function, the above scheme retains weak compressibility.

1.8 Flow with Heat Transfer

The above formulations assumed an isothermal flow field but the LBM can also handle flows with heat transfer. For this purpose, we must introduce a new velocity distribution function $g_i(x, t)$ of a temperature field.[18] The two velocity distribution functions f_i and g_i can then be calculated simultaneously.

Similarly to Eq. (1.14), the time evolution of the velocity distribution function $g_i(x, t)$ is given by

$$g_i(x + c_i \Delta x, t + \Delta t) = g_i(x, t) - \frac{1}{\tau_g}[g_i(x, t) - g_i^{eq}(x, t)]. \tag{1.61}$$

The temperature T is obtained by summing the velocity distribution functions g_i:

$$T = \sum_{i=1}^{N} g_i, \tag{1.62}$$

and the local equilibrium distribution function g_i^{eq} is given by [67]

$$g_i^{eq} = E_i T (1 + 3 c_{i\alpha} u_\alpha), \tag{1.63}$$

where u_α is the flow velocity obtained from f_i.

Similarly to Sec. 1.5, we expand g_i and T respectively as follows:

$$g_i = E_i [g_i^{(0)} + (\Delta x) g_i^{(1)} + (\Delta x)^2 g_i^{(2)} + (\Delta x)^3 g_i^{(3)} + \cdots], \tag{1.64}$$

$$T = T^{(0)} + (\Delta x) T^{(1)} + (\Delta x)^2 T^{(2)} + (\Delta x)^3 T^{(3)} + \cdots. \tag{1.65}$$

Applying the asymptotic analysis to Eqs. (1.61)–(1.63), we get the following advection–diffusion equation of $T^{(0)}$ under the solvability condition:

$$\frac{\text{Sh}}{\Delta x} \frac{\partial T^{(0)}}{\partial t} + u_\alpha^{(1)} \frac{\partial T^{(0)}}{\partial x_\alpha} = \frac{1}{3} \left(\tau_g - \frac{1}{2} \right) \frac{\partial^2 T^{(0)}}{\partial x_\alpha^2}. \tag{1.66}$$

This expression is the governing equation of $T^{(0)}$. The governing equation of $T^{(1)}$ becomes a homogeneous linear equation. If both the initial and boundary conditions of $T^{(1)}$ are 0; that is, the initial and boundary conditions of T are independent of Δx and are imposed only on the first term $T^{(0)}$, then $T^{(1)} = 0$ in all regions. Equation (1.66) is the advection–diffusion equation of the temperature $T = T^{(0)} + O((\Delta x)^2)$. The temperature conductivity coefficient α is found to be

---

[18] Here, the temperature is considered as a simple scalar quantity that moves with the flow.

$$\alpha = \frac{1}{3}\left(\tau_g - \frac{1}{2}\right)\Delta x. \tag{1.67}$$

The heat flux vector q_α is also calculated as follows:

$$q_\alpha = -k\frac{\partial T}{\partial x_\alpha} = T(u_{T\alpha} - u_\alpha), \tag{1.68}$$

where

$$u_{T\alpha} = \frac{1}{T}\sum_{i=1}^{N} c_{i\alpha}g_i, \tag{1.69}$$

and k is the thermal conductivity coefficient given by

$$k = \frac{1}{3}\tau_g\Delta x. \tag{1.70}$$

In summary, Eqs. (1.61), (1.62), and (1.63) compute the temperature field T satisfying the advection–diffusion equation with relative errors $O((\Delta x)^2)$. As the temperature is a scalar quantity that moves with the flow, the above method can be applied to mass transfer problems. For example, the MRT model proposed in [161] handles anisotropic diffusion in the advection–diffusion equation, and LBM-based combustion calculations give the temperature and concentration changes caused by advective diffusion and chemical reactions [157].

1.9 Equation of an Acoustic Field

Thus far, we have considered slow phenomena on the diffusive time scale. The asymptotic analysis (S-expansion) described in Sec. 1.5 can also be performed on the fast acoustic time scale. The equation of an acoustic field (1.73) is derived as follows. As Sh = 1 on the acoustic time scale, we have $\Delta t = \Delta x$ (see Appendix A), and the equations corresponding to Eqs. (1.36) and (1.37) are respectively given by

$$\frac{\partial \rho^{(1)}}{\partial t} + \frac{\partial u_\alpha^{(1)}}{\partial x_\alpha} = 0, \tag{1.71}$$

$$\frac{\partial u_\alpha^{(1)}}{\partial t} + \frac{1}{3}\frac{\partial \rho^{(1)}}{\partial x_\alpha} = 0. \tag{1.72}$$

Eliminating the terms involving $u_\alpha^{(1)}$ from the above equations, we get

$$\frac{\partial^2 \rho^{(1)}}{\partial t^2} = \frac{1}{3}\frac{\partial^2 \rho^{(1)}}{\partial x_\alpha^2}. \qquad (1.73)$$

The above equation is the wave equation of the density $\rho^{(1)}$. The density (pressure) fluctuation propagates at the speed of sound given by $c_s = \sqrt{1/3}$.

Clearly, sound waves can be calculated on the acoustic time scale in the LBM [11, 150]. If the calculated results are viewed on the acoustic time scale (each time step), we find the behavior of the sound wave. On the other hand, looking at the same results on the diffusive time scale (every several hundred time steps), we can find the behavior of the incompressible viscous fluid flows.

1.10 LBM-derived Method

Being based on the discrete Boltzmann equation (1.3), the LBM is naturally applicable to compressible fluids and microfluids. However, in compressible fluid calculations, the number of fictitious particle velocities must be increased as the Mach number increases (see, e.g. [153]). Similarly, in microfluidic calculations, the number of fictitious particle velocities must be increased with increasing Knudsen number (see, e.g. [83]). In general, when the number of fictitious particle velocities is increased, the basic equation of the LBM cannot be formulated into a perfect advection-type difference equation (1.6), so the finite-difference LBM [147, 148] is often used.

This book considers only incompressible viscous fluid flows at low Mach number and small Knudsen number. Compressible fluid and microfluid flows are not treated.

1.11 Boundary Condition

In the LBM, we must determine the velocity distribution function f_i in the directions to the fluid region on the boundary. Assuming that the normal vector n_α on the boundary points toward the fluid side, we must determine f_i such that $c_{i\alpha}n_\alpha > 0$. For example, suppose that the region $y \geq 0$ is filled with fluid in the two-dimensional nine-velocity model. On the boundary $y = 0$, we must determine f_3, f_6, and f_7 as shown in Fig. 1.3.

In the following, we adopt the two-dimensional nine-velocity model for simplicity, but the same ideas are extendible to the three-dimensional fifteen-velocity model.

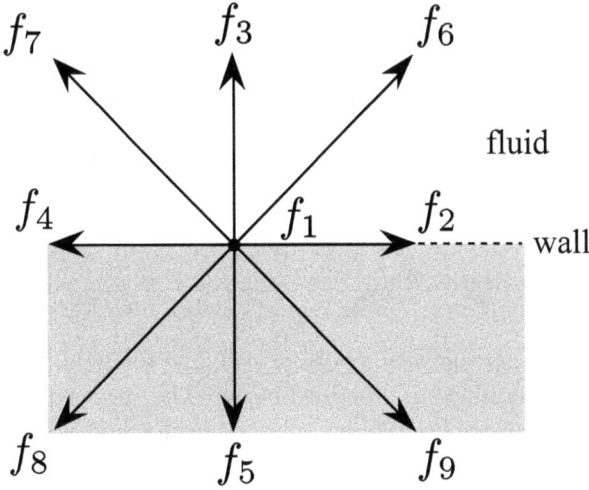

Fig. 1.3 Boundary condition on a plane wall.

(1) No-slip wall
(i) Bounce-back condition

The bounce-back condition is often imposed on no-slip walls. Under the bounce-back condition, fictitious particles encountering the boundary are reflected in the opposite direction to their forward movement. The velocity distribution functions toward the fluid are then determined as follows:

$$\begin{cases} f_3 = f_5 - 6E_5 \; \boldsymbol{c}_5 \cdot \boldsymbol{u}_{\mathrm{w}}, \\ f_6 = f_8 - 6E_8 \; \boldsymbol{c}_8 \cdot \boldsymbol{u}_{\mathrm{w}}, \\ f_7 = f_9 - 6E_9 \; \boldsymbol{c}_9 \cdot \boldsymbol{u}_{\mathrm{w}}, \end{cases} \tag{1.74}$$

where $\boldsymbol{u}_{\mathrm{w}}$ is the moving velocity of the wall. Under the bounce-back condition, the flow velocity perpendicular to the wall is 0, but the flow velocity along the wall is generally non-zero, giving rise to a slip velocity. The slip velocity is known to be very small when $\tau < 1$ [57].

(ii) Counter-slip condition

If you are worried about the slip velocity, you can generate a counter-slip velocity u'_x to impose the no-slip boundary condition [57].

Referring to Eq. (1.17), we first define the unknown velocity distribution functions f_3, f_6, and f_7, respectively:

$$f_3 = \frac{1}{9}\rho'\left\{1 + 3u_{wy} + \frac{9}{2}u_{wy}^2 - \frac{3}{2}\left[(u_{wx} + u_x')^2 + u_{wy}^2\right]\right\}, \tag{1.75}$$

$$f_6 = \frac{1}{36}\rho'\left\{1 + 3(u_{wx} + u_x' + u_{wy}) + \frac{9}{2}(u_{wx} + u_x' + u_{wy})^2 \right.$$
$$\left. - \frac{3}{2}\left[(u_{wx} + u_x')^2 + u_{wy}^2\right]\right\}, \tag{1.76}$$

$$f_7 = \frac{1}{36}\rho'\left\{1 + 3(-u_{wx} - u_x' + u_{wy}) + \frac{9}{2}(-u_{wx} - u_x' + u_{wy})^2 \right.$$
$$\left. - \frac{3}{2}\left[(u_{wx} + u_x')^2 + u_{wy}^2\right]\right\}, \tag{1.77}$$

where $u_{w\alpha}$ is the moving velocity of the wall, and the unknown parameters ρ' and u_x' are determined as described below. The above equation assumes that the velocity distribution function entering the fluid side from the wall is determined by the local equilibrium distribution (1.17) with the counter-slip velocity u_x' along the wall. The density ρ_w on the wall is also unknown. These three unknowns (ρ', u_x', and ρ_w) can be determined by satisfying Eqs. (1.8) and (1.9) at the wall. The solutions are obtained as

$$\rho_w = \frac{1}{1 - u_{wy}}[f_1 + f_2 + f_4 + 2(f_5 + f_8 + f_9)], \tag{1.78}$$

$$\rho' = 6\frac{\rho_w u_{wy} + (f_5 + f_8 + f_9)}{1 + 3u_{wy} + 3u_{wy}^2}, \tag{1.79}$$

$$u_x' = \frac{6}{1 + 3u_{wy}}\frac{\rho_w u_{wx} - (f_2 - f_4 + f_9 - f_8)}{\rho'} - u_{wx}. \tag{1.80}$$

(2) Slip wall (specular condition)

If the fluid slides freely along the wall, the fictitious particles encountering the boundary can be regarded as specularly reflected on the wall. The corresponding boundary condition is

$$\begin{cases} f_3 = f_5, \\ f_6 = f_9, \\ f_7 = f_8. \end{cases} \tag{1.81}$$

(3) Curved wall (improved bounce-back condition)

The above boundary conditions assume a planar boundary that coincides with lattice points. The improved bounce-back condition [164] satisfies the no-slip condition on a curved wall that does not necessarily coincide with lattice points.

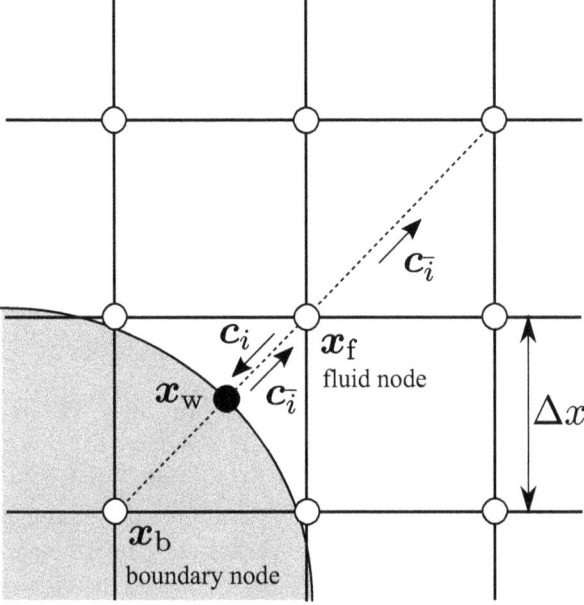

Fig. 1.4   Boundary condition on a curved wall (improved bounce-back condition).

Such a curved wall is shown in Fig. 1.4. We denote two adjacent lattice points across the wall as x_f (fluid side) and x_b (wall side). The particle velocity vector heading from the fluid-side lattice point to the wall-side lattice point is denoted by c_i, and the particle velocity vector heading from the wall side to the fluid side is denoted by $c_{\bar{i}}$ ($= -c_i$). Also, let u_w be the moving velocity of the wall at the boundary point x_w, defined as the intersection of the wall with the line connecting x_f and x_b.

The velocity distribution function $f_{\bar{i}}(x_f)$ toward the fluid side at Point x_f is determined such that the no-slip condition is satisfied at the boundary point x_w. The detailed procedures are shown below.

First, we interpolate the velocity distribution function $f_i(x_w)$ toward the wall side at the boundary point x_w:

$$f_i(x_w) = (1 - \delta) f_i(x_f) + \delta f_i(x_b), \qquad (1.82)$$

where

$$\delta = \frac{|x_f - x_w|}{|x_f - x_b|}. \qquad (1.83)$$

Next, applying the bounce-back condition at the boundary point $\boldsymbol{x}_w$, we determine the velocity distribution function $f_{\bar{i}}(\boldsymbol{x}_w)$ toward the fluid side as follows:

$$f_{\bar{i}}(\boldsymbol{x}_w) = f_i(\boldsymbol{x}_w) - 6E_i \, \boldsymbol{c}_i \cdot \boldsymbol{u}_w. \tag{1.84}$$

Finally, we interpolate to find the velocity distribution function $f_{\bar{i}}(\boldsymbol{x}_f)$ at Point $\boldsymbol{x}_f$ toward the fluid side:

$$f_{\bar{i}}(\boldsymbol{x}_f) = \frac{1}{1+\delta} f_{\bar{i}}(\boldsymbol{x}_w) + \frac{\delta}{1+\delta} f_{\bar{i}}(\boldsymbol{x}_f + \boldsymbol{c}_{\bar{i}}\Delta x). \tag{1.85}$$

(4) Inlet and outlet conditions with a pressure difference

We consider a case (such as a Poiseuille flow) where there is a pressure difference Δp between the inlet ($x = 0$) and the outlet ($x = L$) of the computational domain, and the flow velocities $\boldsymbol{u}$ at the corresponding inlet and outlet points are equal. In this case, referring to the form of the local equilibrium distribution function (1.17), we assume the following forms of the unknown velocity distribution functions $f_2(0)$, $f_6(0)$, and $f_9(0)$ (see Fig. 1.5) [62]:

$$\begin{cases} f_2(0) = f_2(L) + C, \\ f_6(0) = f_6(L) + \frac{1}{4}C, \\ f_9(0) = f_9(L) + \frac{1}{4}C, \end{cases} \tag{1.86}$$

where C is a constant determined as shown below. Meanwhile, the unknown velocity distribution functions $f_4(L)$, $f_7(L)$, and $f_8(L)$ at the outlet (see Fig. 1.5) are assumed to take the forms

$$\begin{cases} f_4(L) = f_4(0) - C, \\ f_7(L) = f_7(0) - \frac{1}{4}C, \\ f_8(L) = f_8(0) - \frac{1}{4}C. \end{cases} \tag{1.87}$$

Finally, using Eq. (1.8) and the relationship $p = \rho/3$, we determine C as the value for which the pressure difference between the inlet and outlet becomes Δp:

$$C = \Delta p - \frac{1}{3}\left[f_1(0) - f_1(L) + f_3(0) - f_3(L) + f_5(0) - f_5(L) \right]. \tag{1.88}$$

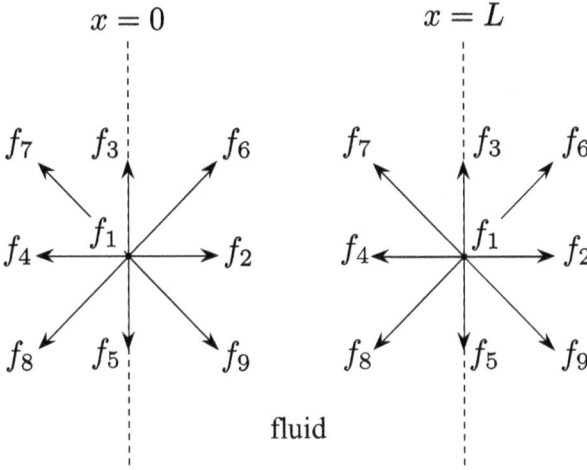

Fig. 1.5 Boundary condition at the inlet ($x = 0$) and outlet ($x = L$) with a pressure difference.

(5) Periodic boundary condition

To impose a periodic boundary condition on the left and right sides of the computational domain ($x = 0$ and $x = L$, respectively), we set $C = 0$ in Eqs. (1.86) and (1.87).

(6) Isothermal wall

We consider a stationary isothermal wall ($T = T_w$) along $y = 0$ as shown in Fig. 1.3. As when imposing the counter-slip condition on the no-slip wall, we refer to the local equilibrium distribution function (1.63) and assume the following forms of the unknown velocity distribution functions g_3, g_6, and g_7 [67, 162]:

$$g_3 = \frac{1}{9}T', \tag{1.89}$$

$$g_6 = \frac{1}{36}T', \tag{1.90}$$

$$g_7 = \frac{1}{36}T'. \tag{1.91}$$

The unknown parameter T' in the above equations is determined by Eq. (1.62) such that $T = T_w$ is satisfied on the wall. The result is

$$T' = 6(T_w - g_1 - g_2 - g_4 - g_5 - g_8 - g_9). \tag{1.92}$$

(7) Isoheat-flux wall

Specifying the heat flux q_y in the y-direction as $q_y = q_w$ on the stationary wall ($y = 0$), we write the unknown velocity distribution functions g_3, g_6, and g_7 as Eqs. (1.89), (1.90), and (1.91), respectively (as in the isothermal wall case). We then determine the unknown parameter T' from Eq. (1.68) for which the heat flux in the y-direction along the wall becomes q_w. The result is

$$T' = 6(q_w + g_5 + g_8 + g_9). \tag{1.93}$$

(8) Heat insulation wall

If the stationary wall along $y = 0$ is a heat insulation wall, we simply set $q_w = 0$ in Eq. (1.93).

1.12　Computational Algorithm of LBM

We now summarize the computational algorithm of the LBM for flows with heat (mass) transfer. In the absence of heat (mass) transfer, the computation of the velocity distribution function g_i can be omitted. In the following, we assume $f_i^{\mathrm{eq,in}}$ given by Eq. (1.58) as the local equilibrium distribution function, but f_i^{eq} given by Eq. (1.17) can also be used.

> **Step 0.** Given the initial values $f_i(\boldsymbol{x}, 0)$ and $g_i(\boldsymbol{x}, 0)$, compute $p(\boldsymbol{x}, 0)$, $\boldsymbol{u}(\boldsymbol{x}, 0)$, and $T(\boldsymbol{x}, 0)$ using Eqs. (1.59), (1.60), and (1.62), respectively.
>
> **Step 1.** Compute the local equilibrium distribution functions $f_i^{\mathrm{eq,in}}$ and g_i^{eq} using Eqs. (1.58) and (1.63), respectively.
>
> **Step 2.** Compute $f_i(\boldsymbol{x}, t + \Delta t)$ and $g_i(\boldsymbol{x}, t + \Delta t)$ using Eqs. (1.57) and (1.61), respectively. Then determine $f_i(\boldsymbol{x}, t + \Delta t)$ and $g_i(\boldsymbol{x}, t + \Delta t)$ entering the fluid side to satisfy the boundary conditions on the boundary of the computational domain.
>
> **Step 3.** Compute $p(\boldsymbol{x}, t + \Delta t)$, $\boldsymbol{u}(\boldsymbol{x}, t + \Delta t)$, and $T(\boldsymbol{x}, t + \Delta t)$ using Eqs. (1.59), (1.60), and (1.62), respectively.
>
> **Step 4.** Advance the time by Δt and return to **Step 1**.

Note that τ and τ_g in Eqs. (1.57) and (1.61) are derived backward from the values of ν and α in Eqs. (1.49) and (1.67), respectively.

As mentioned above, the computational algorithm of the LBM is very simple. In particular, note that it avoids the need to solve the Poisson equation of pressure.

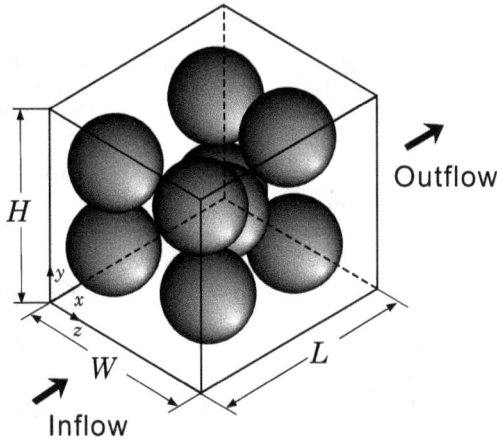

Fig. 1.6   Three-dimensional porous structure.

1.13   Numerical Examples

1.13.1   *Flows in a porous structure*

As an example of flows through complex geometries, we simulate flows in a three-dimensional porous structure [59, 60]; specifically, a rectangular domain containing nine identical spherical bodies (Fig. 1.6). The whole domain is divided into a $73\Delta x \times 69\Delta x \times 69\Delta x$ cubic lattice. Each spherical body is constructed as a lattice block with an equivalent diameter of $D_p = 29.4\Delta x$. The porosity of the structure is $\epsilon = 0.654$. The Reynolds number $\text{Re} = \bar{\rho}_{\text{in}}\bar{u}_{\text{in}}D_p/\mu$ is controlled within the range $0.842 \leq \text{Re} \leq 159$ by changing the pressure difference Δp between the inlet and outlet and the fluid viscosity coefficient μ. Here, $\bar{\rho}_{\text{in}}$ and $\bar{u}_{\text{in}}$ are the time- and space-averaged fluid density and inlet velocity after the transitional flows, respectively.[19] The slip wall condition is applied at the other sides of the domain, and the no-slip (counter-slip) condition is imposed on the bodies. Note that the counter-slip velocity is set to $u'_x = 0$ to avoid numerical instability, which often occurs at high Reynolds numbers.

---

[19] During long-term computations, the sum of the fluid densities in the domain changes because the mass is not completely conserved. This violation is caused by slight numerical errors on the boundary lines and corners (singular points) of the lattice blocks. As $\bar{\rho}_{\text{in}} \neq 1$, we thus replace ν in Eq. (1.49) with the fluid viscosity coefficient μ in order to determine the value of τ.

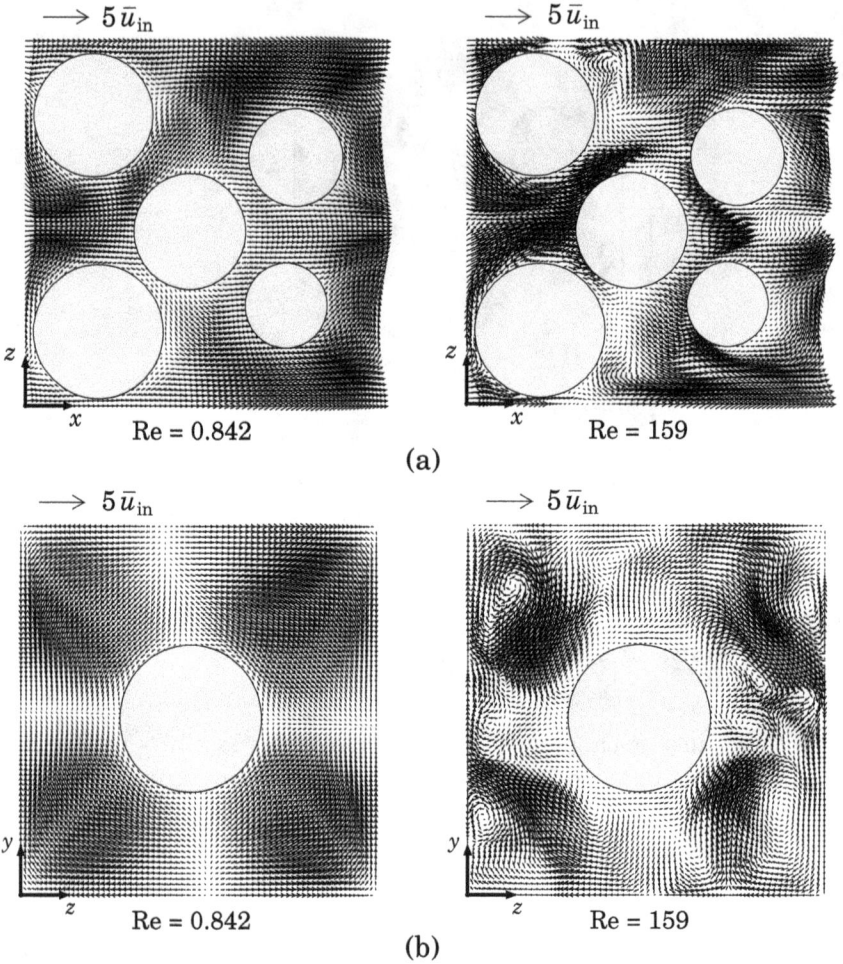

Fig. 1.7 Velocity vectors in a porous structure: (a) $y/H = 0.62$; (b) $x/L = 0.51$. Reprinted from Figs. 5 and 6 in [59] with permission from John Wiley and Sons.

Figure 1.7 shows the velocity vectors calculated on the $y/H = 0.62$ and $x/L = 0.51$ planes at different Reynolds numbers. At the lowest Reynolds number (Re = 0.842), the flow passes through the open spaces while avoiding the bodies. In contrast, at moderate Reynolds number (Re = 159), the flow separates and variously sized vortices appear behind the bodies. In particular, the flow characteristic at $x/L = 0.51$ is quite different from that

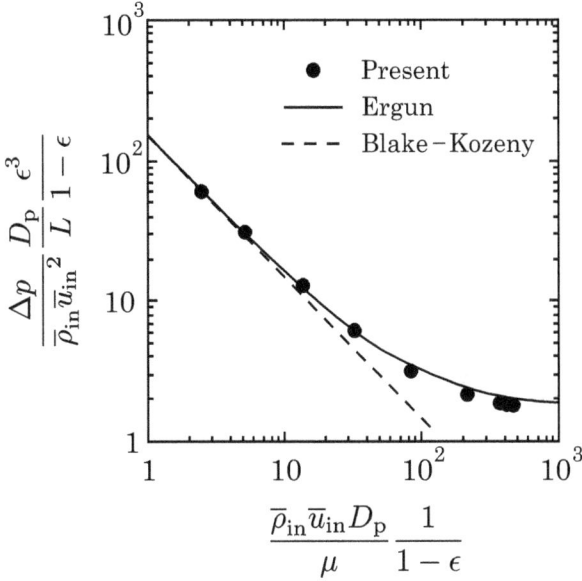

Fig. 1.8 Relationship between pressure drop and Reynolds number. Reprinted from Fig. 8 in [59] with permission from John Wiley and Sons.

at low Reynolds number. Moreover, at the moderate Reynolds number, the strength of each vortex and the flow field itself vary over time.

Figure 1.8 shows the relationship between the dimensionless pressure drop and Reynolds number $Re/(1 - \epsilon)$. This plot compares the computational results with those of the existing empirical equations, namely, the Blake–Kozeny equation [8] (dashed lines) and the Ergun equation [27] (solid lines). The calculated pressure drops clearly agree with those of the empirical equations at low and moderate Reynolds numbers. We also simulated the mass transfer of a binary miscible fluid mixture in the same porous structure using the method described in Sec. 1.8 [162].

1.13.2 *Rayleigh–Bénard convection*

As a typical example of heat transfer, we simulate Rayleigh–Bénard convection [67] in a two-dimensional rectangular domain of width $2H$ and height H. The domain is filled with an incompressible viscous fluid. The temperature T_L on the lower wall at $y = 0$ is maintained higher than the temperature T_U on the upper wall at $y = 1$ ($T_L > T_U$), and periodic

(a)

(b)

(c)

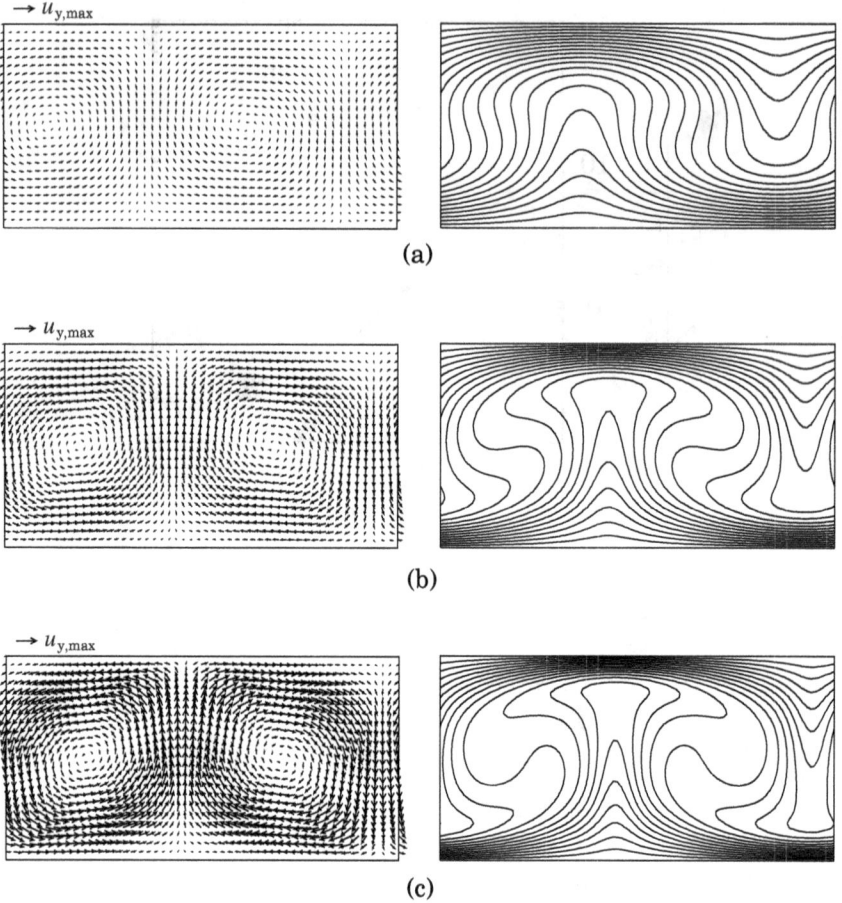

Fig. 1.9   Calculated velocity vectors and isotherms in the simulation of Rayleigh–Bénard convection: (a) Ra = 5000; (b) Ra = 20000; (c) Ra = 50000, where $u_{y,\max}$ is the maximum velocity in the y-direction at Ra = 50000. Reprinted from Fig. 4 in [67] with permission from Elsevier.

boundaries are assumed in the lateral direction. The buoyancy force induced by thermal fluid expansion of the fluid is treated by the Boussinesq approximation. Specifically, we set $G_\alpha(\boldsymbol{x}, t) = g\beta[T(\boldsymbol{x}, t) - T^*]\delta_{\alpha y}$ in Eq. (1.51), where g is the gravitational acceleration, β is the coefficient of thermal expansion, $T^* = T_{\mathrm{L}} - (T_{\mathrm{L}} - T_{\mathrm{U}})y/H$ is the reference temperature, and $\delta_{\alpha y}$ is the Kronecker delta. The whole domain is divided into a $100\Delta x \times 50\Delta x$ square lattice. The dimensionless parameters in this

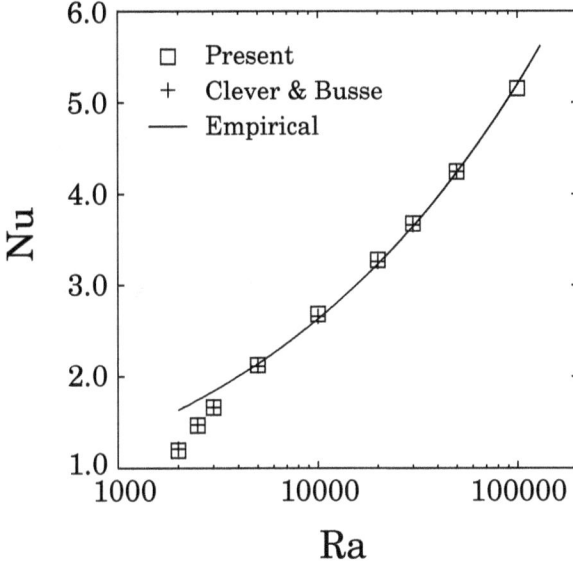

Fig. 1.10 Relationship between the Nusselt and Rayleigh numbers. Reprinted from Fig. 5 in [67] with permission from Elsevier.

problem are the Prandtl number Pr and the Rayleigh number Ra, respectively defined as follows:

$$Pr = \frac{\nu}{\alpha},$$ (1.94)

$$Ra = \frac{g\beta(T_L - T_U)H^3}{\nu^2}Pr.$$ (1.95)

For fixed Prandtl number (Pr = 0.71), the Rayleigh number is ranged from 1650 to 100000. We first consider the critical Rayleigh number Ra_c at which convection occurs. At Rayleigh numbers around 1700, we observe the growth or decay rate of a small perturbation applied to the initial temperature field. The critical Rayleigh number Ra_c is then determined as the Rayleigh number corresponding to zero growth rate. From the calculated results we obtain $Ra_c = 1708.48$, which agrees within 0.042% of the theoretical value (1707.76) obtained by the linear stability theory.

At higher Rayleigh numbers, we then calculate the Nusselt number and compare the results with the existing results to validate the simulation. The Nusselt number is calculated as

$$\text{Nu} = 1 + \frac{\langle u_y T \rangle}{\alpha(T_\text{L} - T_\text{U})/H},\tag{1.96}$$

where $\langle \, \cdot \, \rangle$ denotes the average over the whole domain. The calculated velocity vectors and isotherms are plotted in Fig. 1.9. Increasing the Rayleigh number enhances both the mixing of the hot and cold fluids, and the temperature gradients near the lower and upper walls become steeper. Figure 1.10 relates the Nusselt number to the Rayleigh number. The present and existing results clearly agree [20,46], confirming the sufficient numerical accuracy of the LBM.

Chapter 2

Lattice Kinetic Scheme

As explained in Chap. 1, the LBM is a numerical method that computes incompressible viscous fluids with heat transfer to second-order spatial accuracy. However, because it uses velocity distribution functions, the LBM requires more memory than the usual codes of computational fluid dynamics. This disadvantage can be mitigated by the lattice kinetic scheme [66] (hereafter called the LKS) described below. The LKS can be regarded as an extension of the LBM. In addition, the higher-order dissipation error, which increases when calculating high Reynolds number flows, can be suppressed by applying the link-wise artificial compressibility method (LWACM) [5] or the improved LKS [135].

This chapter explains the LKS, LWACM, and improved LKS. In numerical examples, it then compares the numerical accuracies of these schemes with that of the LBM. The LKS and improved LKS are also demonstrated later in two-phase flow simulations (Chap. 4).

2.1 LKS

We first explain the LKS. Setting $\tau = 1$ and replacing $\boldsymbol{x}$ by $(\boldsymbol{x} - \boldsymbol{c}_i \Delta x)$ in Eq. (1.14), we get

$$f_i(\boldsymbol{x}, t + \Delta t) = f_i^{\text{eq}}(\boldsymbol{x} - \boldsymbol{c}_i \Delta x, t). \tag{2.1}$$

The macroscopic variables can then be calculated as follows:

$$\rho(\boldsymbol{x}, t + \Delta t) = \sum_{i=1}^{N} f_i^{\text{eq}}(\boldsymbol{x} - \boldsymbol{c}_i \Delta x, t), \tag{2.2}$$

$$\rho(\boldsymbol{x}, t + \Delta t)\boldsymbol{u}(\boldsymbol{x}, t + \Delta t) = \sum_{i=1}^{N} \boldsymbol{c}_i f_i^{\text{eq}}(\boldsymbol{x} - \boldsymbol{c}_i \Delta x, t). \tag{2.3}$$

From Eq. (1.17), we find that the local equilibrium distribution function f_i^{eq} on the right-hand side of Eqs. (2.2) and (2.3) is determined only by c_i and the macroscopic variables ρ and u. Therefore, Eqs. (2.2) and (2.3) give the macroscopic variables ρ and u without computing the velocity distribution function f_i. The pressure can be calculated as $p = \rho/3$. In this scheme, the kinematic viscosity coefficient ν given by Eq. (1.49) becomes

$$\nu = \frac{1}{6}\Delta x. \tag{2.4}$$

Although no problems appear at first glance, this scheme cannot efficiently compute high Reynolds number flows. For example, consider a two-dimensional Poiseuille flow between parallel walls with spacing H. If H is divided into $50\Delta x$ and the flow velocity is $u = 0.1$, the Reynolds number becomes $\mathrm{Re} = uH/\nu = 30$, which is rather small. Of course, increasing the number of divisions increases the Reynolds number and the computational efficiency becomes very poor. By setting τ close to 0.5 in the LBM, we can reduce the kinematic viscosity coefficient ν without changing the number of divisions (see Eq. (1.49)).

Thus, it should be possible to change the kinematic viscosity coefficient without changing the number of divisions in the LKS. As shown in Chap. 1, only τ and f_i^{eq} can be changed in the LBM. In the LKS, $\tau = 1$ is fixed so only f_i^{eq} can be changed. The Chapman–Enskog velocity distribution function is a well-known local equilibrium distribution function for the Navier–Stokes equations in the general kinetic theory of gases [19, 77, 127]. Accordingly, we consider the following local equilibrium distribution function of the Chapman–Enskog type in the LKS:

$$\rho(\boldsymbol{x}, t + \Delta t) = \sum_{i=1}^{N} f_i^{eq,\mathrm{LKS}}(\boldsymbol{x} - \boldsymbol{c}_i \Delta x, t), \tag{2.5}$$

$$\rho(\boldsymbol{x}, t + \Delta t)\boldsymbol{u}(\boldsymbol{x}, t + \Delta t) = \sum_{i=1}^{N} \boldsymbol{c}_i f_i^{eq,\mathrm{LKS}}(\boldsymbol{x} - \boldsymbol{c}_i \Delta x, t), \tag{2.6}$$

where

$$f_i^{eq,\mathrm{LKS}} = f_i^{eq} + A\Delta x E_i \rho \left(\frac{\partial u_\beta}{\partial x_\alpha} + \frac{\partial u_\alpha}{\partial x_\beta} \right) c_{i\alpha} c_{i\beta}. \tag{2.7}$$

The second term in the above equation is an additional term for adjusting the viscosity coefficient, and $A = O(1)$ is a constant parameter. In the presence of an external force, we must add the term $3\Delta x E_i c_{i\alpha} G_\alpha$ (where

G_α is the external force acting per unit mass) to the right-hand side of Eq. (2.7), as discussed in Sec. 1.6.

If the second term on the right-hand side of the local equilibrium distribution function in Eq. (2.7) vanishes, the LKS (like the LBM) excellently conserves mass and momentum. When the second term on the right-hand side is included, the LKS includes relative errors $O((\Delta x)^2)$ in the conservations of mass and momentum.

The fluid temperature T can be similarly obtained without using the velocity distribution function g_i by

$$T(\boldsymbol{x}, t + \Delta t) = \sum_{i=1}^{N} g_i^{\text{eq,LKS}}(\boldsymbol{x} - \boldsymbol{c}_i \Delta x, t). \tag{2.8}$$

Here, $g_i^{\text{eq,LKS}}$ is the local equilibrium distribution function given by

$$g_i^{\text{eq,LKS}} = g_i^{\text{eq}} + E_i B \Delta x c_{i\alpha} \frac{\partial T}{\partial x_\alpha}, \tag{2.9}$$

where $B = O(1)$ is a constant parameter that determines the temperature conductivity coefficient.

The first-order derivatives in Eqs. (2.7) and (2.9) can be approximated as

$$\frac{\partial \phi}{\partial x_\alpha} \approx \frac{1}{6\Delta x} \sum_{i=1}^{9} c_{i\alpha} \phi(\boldsymbol{x} + \boldsymbol{c}_i \Delta x), \tag{2.10}$$

in the two-dimensional nine-velocity model and

$$\frac{\partial \phi}{\partial x_\alpha} \approx \frac{1}{10\Delta x} \sum_{i=1}^{15} c_{i\alpha} \phi(\boldsymbol{x} + \boldsymbol{c}_i \Delta x), \tag{2.11}$$

in the three-dimensional fifteen-velocity model.

Substituting the expanded forms of Eqs. (1.21), (1.22), and (1.65) into the Taylor-expanded forms of Eqs. (2.5), (2.6), and (2.8) (expanded around $(\boldsymbol{x}, t)$), and collecting terms with the same order of Δx on the diffusive time scale ($\Delta t = \text{Sh}\Delta x = O((\Delta x)^2)$), we get Eqs. (1.44), (1.45), and (1.66). In addition, we can respectively write the kinematic viscosity coefficient ν, the temperature conductivity coefficient α, and the thermal conductivity coefficient k as

$$\nu = \left(\frac{1}{6} - \frac{2}{9} A \right) \Delta x, \qquad (2.12)$$

$$\alpha = \left(\frac{1}{6} - \frac{1}{3} B \right) \Delta x, \qquad (2.13)$$

$$k = \left(\frac{1}{3} - \frac{1}{3} B \right) \Delta x. \qquad (2.14)$$

Thus, the LKS computes the continuity equation, the Navier–Stokes equations, and the temperature advection–diffusion equation for incompressible viscous fluids to second-order spatial accuracy. We can arbitrarily change the Reynolds number Re = uL/ν and Prandtl number Pr = ν/α by appropriately choosing A and B. Following the treatment in Sec. 1.9, the speed of sound in the LKS becomes $c_s = \frac{1}{\sqrt{3}} \sqrt{1 - \frac{2}{3} A}$.

In calculations of flows with high Reynolds numbers, the LKS is known to be numerically more stable than the LBM because the LKS includes higher-order error terms of the fourth-order spatial derivative of u (known as the dissipation error due to hyper-viscosity). As revealed in the above Taylor expansion of Eq. (2.6), the viscous term of the Navier–Stokes equations (the second-order spatial derivative of u) is the sum of the part derived from f_i^{eq} (the first term of Eq. (2.7)) and the part derived from the second term of Eq. (2.7). The first and second parts contribute the terms $\frac{1}{6} \Delta x$ and $-\frac{2}{9} A \Delta x$, respectively, to the kinematic viscosity coefficient given by Eq. (2.12). When calculating high Reynolds number flows, the value of $\left(\frac{1}{6} - \frac{2}{9} A \right)$ needs to be small, but the dissipation error due to hyper-viscosity is usually not proportional to $\left(\frac{1}{6} - \frac{2}{9} A \right)$.[1] Therefore, if $\left(\frac{1}{6} - \frac{2}{9} A \right)$ becomes too small, the viscous term becomes of the same order as the dissipation error due to hyper-viscosity. In that case, although the computation is stable, the numerical viscosity might increase.[2]

Recently, the numerical stability of the LKS has been studied in detail [53].

2.2 LWACM

When calculating high Reynolds number flows, the dissipation error in the LKS can be reduced by the LWACM [5], which is formulated as follows:

---

[1] Fortunately, the hyper-viscosity is proportional to $(\tau - \frac{1}{2})$ in the LBM.

[2] The problem of the dissipation error due to hyper-viscosity has been pointed out in a study using the MRT model for the collision term [40].

$$p(\boldsymbol{x}, t + \Delta t) = \frac{1}{3} \sum_{i=1}^{N} \Big\{ f_i^{\text{eq,in}}(\boldsymbol{x} - \boldsymbol{c}_i \Delta x, t)$$

$$+ 3 A_p E_i \boldsymbol{c}_i \cdot [\boldsymbol{u}(\boldsymbol{x}, t) - \boldsymbol{u}(\boldsymbol{x} - \boldsymbol{c}_i \Delta x, t)] \Big\}, \qquad (2.15)$$

$$\boldsymbol{u}(\boldsymbol{x}, t + \Delta t) = \sum_{i=1}^{N} \boldsymbol{c}_i \Big\{ f_i^{\text{eq,in}}(\boldsymbol{x} - \boldsymbol{c}_i \Delta x, t)$$

$$+ 3 A_u E_i \boldsymbol{c}_i \cdot [\boldsymbol{u}(\boldsymbol{x}, t) - \boldsymbol{u}(\boldsymbol{x} - \boldsymbol{c}_i \Delta x, t)] \Big\}, \qquad (2.16)$$

where $f_i^{\text{eq,in}}$ is the local equilibrium distribution function defined by Eq. (1.58), and $A_p = O(1)$ and $A_u = O(1)$ are constants controlling the speed of sound and the viscosity coefficient, respectively.

Substituting the expanded forms of Eqs. (1.22) and (1.24) into the Taylor-expanded forms of Eqs. (2.15) and (2.16) (expanded around $(\boldsymbol{x}, t)$), and collecting terms with the same order of Δx, we get Eqs. (1.44) and (1.45). In addition, the kinematic viscosity coefficient ν is given by

$$\nu = \frac{1}{6} (1 - A_u) \Delta x. \qquad (2.17)$$

As shown in the above asymptotic analysis, a hyper-viscosity-related dissipation error occurs in the LWACM similarly to the LKS, but its magnitude decreases proportionally with decreasing $(1 - A_u)$ and ν. Therefore, when calculating high Reynolds number flows, the dissipation error remains small because $(1 - A_u)$ is small, so the dissipation error due to hyper-viscosity can be suppressed. However, the speed of sound in the LWACM is $c_s = \frac{1}{\sqrt{3}} \sqrt{1 - A_p}$. The LWACM is destabilized when the speed of sound approaches $1/\sqrt{3}$. Therefore, excessive numerical oscillations caused by sound waves might increase the acoustic error in the LWACM (see Sec. 2.7.2 for a numerical example).

2.3 Improved LKS

The improved LKS [135] suppresses both the dissipation error due to hyper-viscosity and the acoustic error due to excessive numerical oscillations caused by sound waves. In this scheme, the dissipation errors are reduced by applying the LWACM, and the acoustic errors are reduced by internally iterating the pressure calculation and implicitly introducing the pressure result into the flow velocity computation.

The improved LKS first determines the pressure iteratively as follows:

$$p_{l+1}(\boldsymbol{x}) = \frac{1}{3} \sum_{i=1}^{N} \left\{ f_i^{\text{eq,in}} (p_l(\boldsymbol{x} - \boldsymbol{c}_i \Delta x), \boldsymbol{u}(\boldsymbol{x} - \boldsymbol{c}_i \Delta x, t)) \right.$$

$$\left. + 3A_p E_i \boldsymbol{c}_i \cdot [\boldsymbol{u}(\boldsymbol{x}, t) - \boldsymbol{u}(\boldsymbol{x} - \boldsymbol{c}_i \Delta x, t)] \right\}, \qquad (2.18)$$

where $f_i^{\text{eq,in}}$ is the local equilibrium distribution function defined by Eq. (1.58). Note that $p_0(\boldsymbol{x}) = p(\boldsymbol{x}, t)$. This iteration is repeated n times. As shown in the following numerical examples, a sufficient condition is $n \leq 5$. After n iterations, the pressure at the new time step is updated as $p(\boldsymbol{x}, t + \Delta t) = p_n(\boldsymbol{x})$.

Next, the flow velocity $\boldsymbol{u}(\boldsymbol{x}, t + \Delta t)$ is computed as

$$\boldsymbol{u}(\boldsymbol{x}, t + \Delta t) = \sum_{i=1}^{N} \boldsymbol{c}_i \left\{ f_i^{\text{eq,in}} (p(\boldsymbol{x} - \boldsymbol{c}_i \Delta x, t + \Delta t), \boldsymbol{u}(\boldsymbol{x} - \boldsymbol{c}_i \Delta x, t)) \right.$$

$$\left. + 3A_u E_i \boldsymbol{c}_i \cdot [\boldsymbol{u}(\boldsymbol{x}, t) - \boldsymbol{u}(\boldsymbol{x} - \boldsymbol{c}_i \Delta x, t)] \right\}, \qquad (2.19)$$

where $p(\boldsymbol{x}, t + \Delta t)$ is the pressure at the new time step. In Sec. 2.4, we will show that using $p(\boldsymbol{x}, t + \Delta t)$ increases the bulk viscosity coefficient.

The governing equations of p and $\boldsymbol{u}$ in the above computational scheme can be obtained through the following asymptotic analysis (S-expansion). The expanded forms of Eqs. (2.18) and (2.19) are substituted into the Taylor-expanded forms of Eqs. (2.15) and (2.16) around $(\boldsymbol{x}, t)$. Collecting terms with the same order of Δx on the diffusive time scale ($\Delta t = \text{Sh}\Delta x = O((\Delta x)^2)$), we get the following equations:

$$\frac{\partial u_\alpha^{(1)}}{\partial x_\alpha} = 0, \qquad (2.20)$$

$$\frac{\text{Sh}}{\Delta x} \frac{\partial u_\alpha^{(1)}}{\partial t} + u_\beta^{(1)} \frac{\partial u_\alpha^{(1)}}{\partial x_\beta} = -\frac{\partial p^{(2)}}{\partial x_\alpha} + \frac{1}{6}(1 - A_u) \frac{\partial^2 u_\alpha^{(1)}}{\partial x_\beta^2}. \qquad (2.21)$$

Note that $p^{(1)}$ is constant on the diffusive time scale. Under appropriate initial and boundary conditions, $u_\alpha^{(2)} = p^{(3)} = 0$ in all regions. Therefore, $u_\alpha = (\Delta x) u_\alpha^{(1)} + O((\Delta x)^3)$ and $p = 1/3 + (\Delta x)^2 p^{(2)} + O((\Delta x)^4)$ satisfy the continuity equation (2.20) and the Navier–Stokes equations (2.21) for incompressible viscous fluids with relative errors $O((\Delta x)^2)$. The kinematic viscosity coefficient ν is the same as in the LWACM, namely,

$$\nu = \frac{1}{6}(1 - A_u)\,\Delta x. \tag{2.22}$$

Meanwhile, the following equations hold on the acoustic time scale ($\Delta t = \Delta x$):

$$\frac{\partial p^{(1)}}{\partial t} + \frac{n}{3}(1 - A_p)\frac{\partial u_\alpha^{(1)}}{\partial x_\alpha} = 0, \tag{2.23}$$

$$\frac{\partial u_\alpha^{(1)}}{\partial t} + \frac{\partial p^{(1)}}{\partial x_\alpha} = 0. \tag{2.24}$$

Eliminating the terms involving $u_\alpha^{(1)}$ from both equations, we get

$$\frac{\partial^2 p^{(1)}}{\partial t^2} = \frac{n}{3}(1 - A_p)\frac{\partial^2 p^{(1)}}{\partial x_\alpha^2}. \tag{2.25}$$

Note that the speed of sound in the improved LKS becomes $c_{\mathrm{s}} = \sqrt{\frac{n}{3}}\sqrt{1 - A_p}$.

2.4 Bulk Viscosity Coefficient

The above-described schemes have weak compressibility, meaning that bulk viscosity plays an important role in their numerical stability.

As revealed in the asymptotic analyses (S-expansion) of Secs. 1.5 and 2.3, $\partial u_\alpha^{(1)}/\partial x_\alpha = \partial u_\alpha^{(2)}/\partial x_\alpha = 0$ in both the LBM and LKS. That is, the compressibility effects (including the bulk viscosity) appear in relative errors $O((\Delta x)^2)$. Under the solvability condition (1.35) for $m = 5$, we can derive an equation concerning $\partial(\rho^{(2)}u_\alpha^{(1)} + u_\alpha^{(3)})/\partial t$. From the coefficient of $\partial^2(\rho^{(2)}u_\beta^{(1)} + u_\beta^{(3)})/\partial x_\alpha \partial x_\beta$ in the derived equation, we find that $\zeta = \nu$, where ζ is the bulk viscosity coefficient (the derivation is omitted). This ζ does not affect the results calculated to second-order spatial accuracy, but does affect the numerical stability. It is known that when ζ is more than 10 times larger than ν, the acoustic error due to excessive numerical oscillations caused by sound waves is suppressed and the numerical stability is improved [25, 111]. In fact, when the fluid is expanded ($\nabla \cdot \boldsymbol{u} > 0$) and compressed ($\nabla \cdot \boldsymbol{u} < 0$), the effective pressure p_{eff} ($= p - \zeta \nabla \cdot \boldsymbol{u}$) decreases

Table 2.1 Sound speeds c_s, kinematic viscosity coefficients ν, and bulk viscosity coefficients ζ and ζ' in different computational schemes. Reprinted from Table 1 in [142] with permission from AIP Publishing.

	LBM	LKS	LWACM	improved LKS
c_s	$\frac{1}{\sqrt{3}}$	$\frac{1}{\sqrt{3}}\sqrt{1-\frac{2}{3}A}$	$\frac{1}{\sqrt{3}}\sqrt{1-A_p}$	$\sqrt{\frac{n}{3}}\sqrt{1-A_p}$
ν	$\frac{1}{3}\left(\tau-\frac{1}{2}\right)\Delta x$	$\left(\frac{1}{6}-\frac{2}{9}A\right)\Delta x$	$\frac{1}{6}\left(1-A_u\right)\Delta x$	$\frac{1}{6}\left(1-A_u\right)\Delta x$
ζ	ν	$2\nu-\frac{1}{2}c_s^2\Delta x$	$2\nu-\frac{1}{2}c_s^2\Delta x$	$2\nu+\frac{1}{2}c_s^2\Delta x$
ζ'	2ν	2ν	2ν	$2\nu+c_s^2\Delta x$

and increases, respectively, and the pressure oscillations are suppressed. When numerical oscillations caused by sound waves occur in calculations, $\partial u_\alpha^{(2)}/\partial x_\alpha$ can be non-zero. As seen in Eq. (1.43), the bulk viscosity coefficient is not ζ but $\zeta' = 2\nu$, defining the bulk viscosity coefficient when $\partial u_\alpha^{(2)}/\partial x_\alpha \neq 0$.

In particular, when calculating high Reynolds number flows on a low-resolution grid (on which Δx cannot be reduced), we must reduce ν by fixing τ close to 0.5 (see Eq. (1.49)). However, when τ approaches 0.5, the computation becomes unstable as explained in Sec. 1.4. To make the computation stable, we can set ζ much larger than ν ($\zeta \gg \nu$). Table 2.1 gives the bulk viscosity coefficients ζ and ζ', the speed of sound c_s, and the kinematic viscosity coefficient ν in each computational scheme. In the LBM, LKS, and LWACM, ζ and ζ' cannot be much larger than ν,[3] but in the improved LKS, both ζ and ζ' can be much larger than ν.[4] Therefore, the improved LKS is thought to suppress the acoustic error and guarantee excellent numerical stability (this expectation is confirmed in the numerical example in Sec. 2.7). Nevertheless, an excessively large ζ might reduce the numerical accuracy, so verifying the accuracy of the calculated result is essential.

---

[3]When ν is small, ζ can be negative in the LKS and LWACM. At first glance, the computation seems likely to fail as the acoustic error increases uncontrollably. However, if sound waves cause excessive numerical oscillations, $\partial u_\alpha^{(2)}/\partial x_\alpha \neq 0$ and the effect of $\zeta' > 0$ dominates. Thus, the computation may continue even if ζ becomes negative.

[4]The MRT-LBM described in Appendix C also improves the numerical stability by enforcing ζ larger than ν, which suppresses the acoustic error.

2.5 Boundary Condition

The macroscopic variables u, p, and T in the LKS, LWACM, and improved LKS require appropriately specified boundary conditions. The boundary conditions used in conventional methods (such as the finite-difference and finite element methods) are suitable for this purpose. However, as seen in Eqs. (2.7) and (2.9), the first-order derivatives of u and T are required on the boundary. These derivatives can be calculated to second-order accuracy by a suitable scheme such as the one-sided difference approximation.

2.6 Computational Algorithm of Improved LKS

The computational algorithm of the improved LKS is summarized below.

Step 0. Give the initial values of $p(\boldsymbol{x},0)$ and $\boldsymbol{u}(\boldsymbol{x},0)$.
Step 1. Determine the boundary values of $p(\boldsymbol{x}_\mathrm{w},t)$ and $\boldsymbol{u}(\boldsymbol{x}_\mathrm{w},t)$ under the specified boundary condition.
Step 2. Compute the pressure $p(\boldsymbol{x},t+\Delta t)$ as follows:

(2-0) Give the initial value of $p_0(\boldsymbol{x}) = p(\boldsymbol{x},t)$ in the iteration.
(2-1) Compute the pressure in the $(l+1)$th iteration by Eq. (2.18). Perform n iterations of this process.
(2-2) Update the pressure at the new time as $p(\boldsymbol{x},t+\Delta t) = p_n(\boldsymbol{x})$.

Step 3. Compute the flow velocity $\boldsymbol{u}(\boldsymbol{x},t+\Delta t)$ by Eq. (2.19).
Step 4. Advance the time by Δt and return to **Step 1**.

2.7 Numerical Examples

2.7.1 *Natural convection flows in a square cavity*

We first examine the accuracy of the LKS by simulating natural convection flows in a square cavity with adiabatic top and bottom walls. The side walls are maintained at constant but different temperatures as shown in Fig. 2.1. The temperature is assumed higher at the right-hand side wall T_h than at the left-hand side wall T_c. The buoyancy force due to thermal expansion of the fluid is modeled by the Boussinesq approximation. As described in Sec. 1.13.2, the term $3E_i g\beta(T - T^*)c_{iy}\Delta x$ is added to Eq. (2.7), where $T^* = (T_\mathrm{c}+T_\mathrm{h})/2$ is the reference temperature. The dimensionless parameters of this problem are the Prandtl number and the Rayleigh number defined

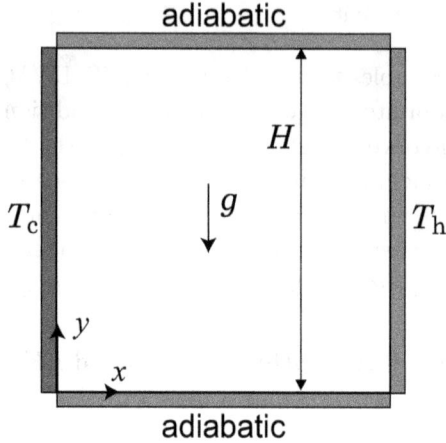

Fig. 2.1   Geometry of a natural convection problem.

by Eqs. (1.94) and (1.95), respectively. Note that $(T_L - T_U)$ in Eq. (1.95) is replaced by $(T_h - T_c)$.

Figure 2.2(a) shows the calculated results for Ra = 10^5 and Pr = 0.1. The computational domain is divided into a $200\Delta x \times 200\Delta x$ square lattice. The computational conditions are $T_c = 1$, $T_h = 2$, $A = 0.74598$, $B = 0.73658$, and $g\beta\Delta x = 10^{-7}$. Natural convection induces a counter-clockwise rotational flow and the isotherms are distorted by the flow field, which alters the temperature field. The accuracy of the result is checked by examining the mean Nusselt number defined as

$$\text{Nu} = \frac{\bar{q}}{\bar{q}_c}, \tag{2.26}$$

where $\bar{q}$ is the actual heat flux across the cavity and $\bar{q}_c$ is the heat flux in pure conduction without flows. The present result (Nu = 3.95) agrees within 1% of Nu = 3.9248 obtained by Ferziger and Perić [34] using the finite-difference method. The velocity vectors and isotherms are also consistent with their results.

We next show the calculated results for Ra = 10^5 and Pr = 0.71 on a $160\Delta x \times 160\Delta x$ square lattice (Fig. 2.2(b)). The computational conditions are $T_c = 1$, $T_h = 2$, $A = 0.67326$, $B = 0.71397$, and $g\beta\Delta x = 10^{-5}$. In this case, two counter-clockwise rotational flows appear in the cavity, and the temperature gradients near the top left and bottom right corners are

(a)

(b)

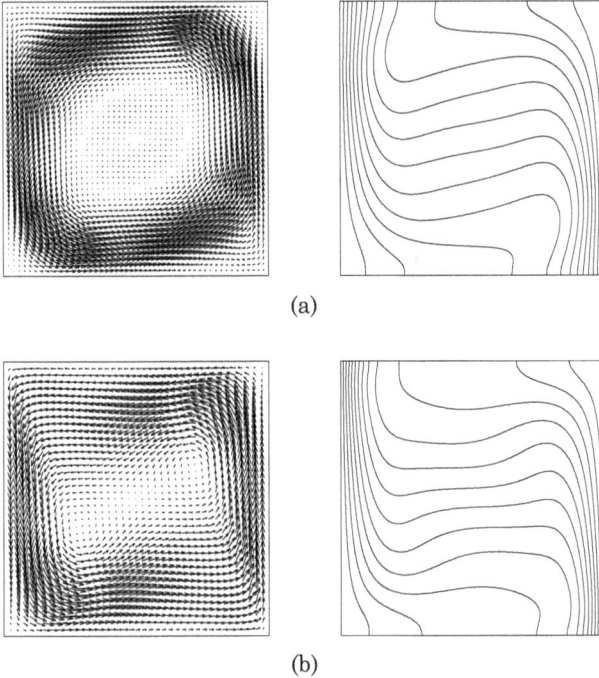

Fig. 2.2 Calculated velocity vectors and isotherms: (a) Ra = 10^5, Pr = 0.1; (b) Ra = 10^5, Pr = 0.71. Reprinted from Figs. 2 and 3 in [66] with permission from The Royal Society (U.K.).

steeper than those in Fig. 2.2(a). The mean Nusselt number (Nu = 4.60) is larger than in the case with Pr = 0.1, and agrees within 1% of Nu = 4.5275 obtained by Hortmann et al. [52] using the finite-volume method.

2.7.2 *Taylor–Green vortex*

This subsection compares the numerical accuracies of the LBM, LKS, LWACM, and improved LKS in a Taylor–Green vortex problem, which admits a simple analytical solution to the incompressible Navier–Stokes equations. As the analytical solution is known, this problem has been widely used as a benchmark problem. In this simulation, we consider a strongly unsteady problem with an external force that oscillates periodically in space and time. Such a problem is called a *generalized* Taylor–Green vortex problem [102].

The periodically oscillating external force is given by

$$\begin{pmatrix} G_x(x,y,t) \\ G_y(x,y,t) \end{pmatrix} = \frac{U_{\max}^2}{D} \left(\frac{2\pi^2}{\mathrm{Re}} \cos\omega t - \sin\omega t \right)$$
$$\times \begin{pmatrix} \sin(k(x - u_0 t))\cos(k(y - v_0 t)) \\ -\cos(k(x - u_0 t))\sin(k(y - v_0 t)) \end{pmatrix}, \quad (2.27)$$

where $U_{\max}$ is the maximum velocity, D is the width of cellular vortex, Re is the Reynolds number defined as $\mathrm{Re} = U_{\max} D/\nu$, ω is the angular frequency given by $\omega = U_{\max}/D$, k is the wave number given by $k = \pi/D$, and $\boldsymbol{u}_c = (u_0, v_0)^{\mathrm{T}}$ is the convective velocity (where the superscript 'T' denotes the transpose of a vector). Under the external force, the analytical solution of the incompressible Navier–Stokes equations is given by

$$p(x,y,t) = p_0 + \frac{1}{4} U_{\max}^2 \left[\cos(2k(x - u_0 t)) + \cos(2k(y - v_0 t)) \right] \cos^2\omega t, \quad (2.28)$$

$$\begin{pmatrix} u(x,y,t) \\ v(x,y,t) \end{pmatrix} = \begin{pmatrix} u_0 \\ v_0 \end{pmatrix} + \begin{pmatrix} U_{\max} \sin(k(x - u_0 t))\cos(k(y - v_0 t))\cos\omega t \\ -U_{\max} \cos(k(x - u_0 t))\sin(k(y - v_0 t))\cos\omega t \end{pmatrix},$$
$$(2.29)$$

where p_0 is a reference pressure. Figure 2.3 shows the analytical flow field at $t = 0$. Here, the velocity vectors are displayed relative to the convective velocity $(\boldsymbol{u} - \boldsymbol{u}_c)$. Under the external force given by Eq. (2.27), the cellular vortices amplify and decay at an angular frequency ω while being convected at velocity $\boldsymbol{u}_c$.

Here we compare the calculated results of the LBM, LKS, LWACM, and improved LKS with the analytical solution to the above problem. The computational domain is set to $[0, 2D] \times [0, 2D]$ with $D = 40\Delta x$. Periodic boundary conditions are applied at all boundaries of the computational domain. The initial analytical solution is set as the initial computational condition. The convective velocity is set to $(u_0, v_0) = U_{\max}(\cos(\pi/12), \sin(\pi/12))$ with $U_{\max} = 0.005$, and the Reynolds number is set to 100. Under this condition, the speed of sound in the LKS is $c_s = 0.7113/\sqrt{3}$. In the improved LKS, we set $A_p = 0$. The speed of sound is then determined as $c_s = \sqrt{n/3}$, where n is the number of iterations in the pressure computation.

Figure 2.4 shows the numerical deviations of pressure, flow velocity, and velocity divergence from the analytical solution. As the error norm, we apply the L^1 norm (i.e., the mean error over all lattice points). As indicated in Fig. 2.4(a) and (b), the errors in the pressure and divergence

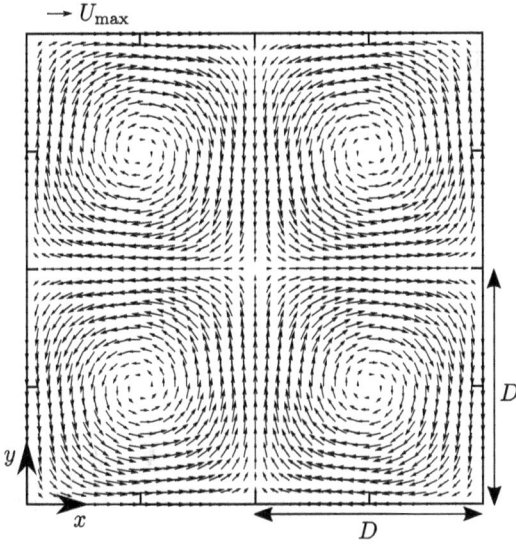

Fig. 2.3 Analytical flow field of a generalized Taylor–Green vortex problem at $t = 0$. The velocity vectors are relative to the convective velocity $(\boldsymbol{u} - \boldsymbol{u}_c)$.

calculated by the LBM, LKS, and LWACM oscillate at high frequency owing to acoustic error. In addition, the error is much larger in the LKS than in the LBM. This difference is caused by the dissipation error due to hyper-viscosity in the LKS. Meanwhile, the error in the LWACM is comparable to that in the LBM, implying that the LWACM generates no significant dissipation error. However, when the speed of sound approaches that in the LBM, $c_s = 1/\sqrt{3}$, the LWACM computation blows up. From Fig. 2.4(b), we observe that decreasing the speed of sound in the LWACM enlarges the divergence error but reduces the pressure error.

On the contrary, the improved LKS eliminates the dissipation error (Fig. 2.4(c)) and diminishes the acoustic error. The acoustic error also decays much faster in the improved LKS than in the other methods. In addition, the error in the velocity divergence is smaller for $n = 5$ than for $n = 1$. Therefore, the improved LKS outperforms the other methods in terms of numerical stability, dissipation error, and acoustic error. The computational time of the improved LKS with $n = 5$ is approximately twice that with $n = 1$.

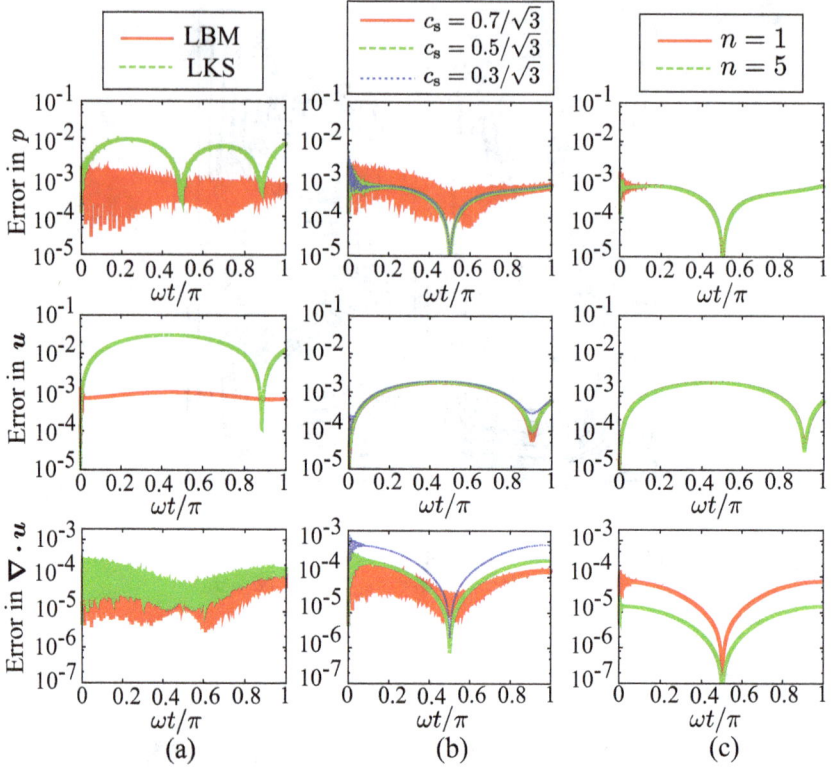

Fig. 2.4 Numerical deviations of pressure p, flow velocity $\boldsymbol{u}$, and velocity divergence $\nabla \cdot \boldsymbol{u}$ from the analytical solution in a generalized Taylor–Green vortex problem: (a) LBM and LKS; (b) LWACM with $c_s = 0.7/\sqrt{3}$, $0.5/\sqrt{3}$, and $0.3/\sqrt{3}$; (c) improved LKS with $n = 1$ and $n = 5$.

2.7.3 *Doubly periodic shear layer*

Finally, we compare the numerical accuracies and stabilities of the LBM, LKS, LWACM, and improved LKS in a simulation of high Reynolds number flows on a low-resolution grid. In this comparison, we simulate a doubly periodic shear layer. Minion and Brown [98] originally examined the accuracies of various numerical methods on this problem.

The initial conditions of the velocity field (corresponding to the perturbed shear layer) are set in a periodic square domain of size $[0, H] \times [0, H]$ (Fig. 2.5):

(a)

(b)

(c)

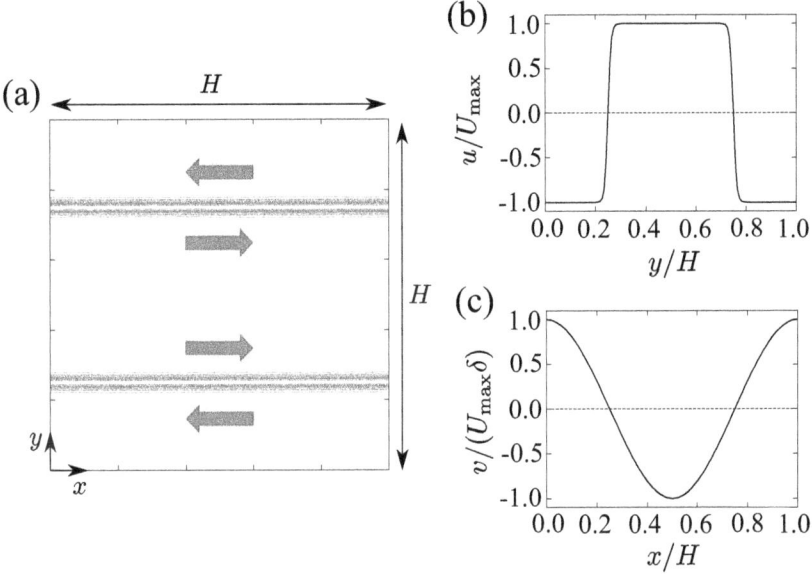

Fig. 2.5   Doubly periodic shear layer at $t = 0$: (a) vorticity contours; (b) distribution of u in the y-direction; (c) distribution of v in the x-direction.

$$\begin{pmatrix} u(x,y,0) \\ v(x,y,0) \end{pmatrix} = \begin{pmatrix} U_{\max} \tanh\left(\alpha(0.25H - |y - 0.5H|)\right) \\ U_{\max}\delta \sin\left(2\pi(x + 0.25H)\right) \end{pmatrix}, \tag{2.30}$$

where $U_{\max}$ is the maximum flow speed, and α and δ are parameters determining the thickness of the shear layer and the amplitude of the initial perturbation in the y-direction, respectively. The initial pressure field is set to satisfy the Navier–Stokes equations with the initial velocity field given by Eq. (2.30). Over time, the shear layers are expected to roll up due to the Kelvin–Helmholtz instability. The governing parameter of this system is the Reynolds number Re $= U_{\max}H/\nu$. Here we set Re $= 10000$, higher than in the above numerical examples. We also set $U_{\max} = 0.03$, $\alpha = 80$, and $\delta = 0.05$. Simulations by Minion and Brown [98] revealed that whereas two large vortices appear on a high-resolution grid ($H = 256\Delta x$), spurious secondary vortices may appear on a low-resolution grid ($H = 128\Delta x$), depending on the numerical method. This problem is a widely employed benchmark problem for testing the numerical stabilities of simulations on low-resolution grids.

Figure 2.6 shows the vorticity contours at $U_{\max}t/H = 1$ on the low-resolution grid ($H = 128\Delta x$) computed by each method. In Fig. 2.6(a),

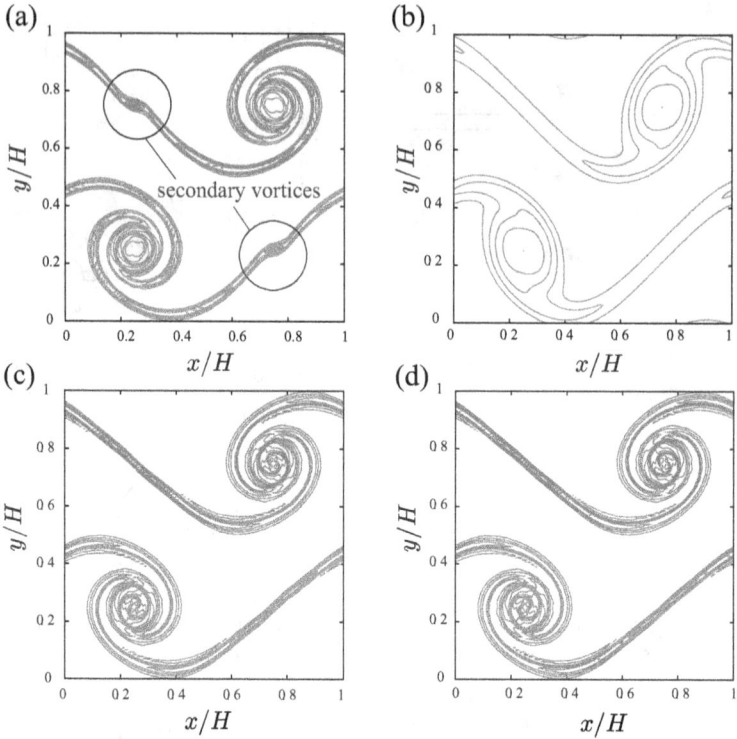

Fig. 2.6   Vorticity contours at $U_{max}t/H = 1$ computed by (a) the LBM, (b) the LKS, (c) the LWACM with $c_s = 0.7/\sqrt{3}$, and (d) the improved LKS with $c_s = 1/\sqrt{3}$ ($n = 1$ and $A_p = 0$). Here, a doubly periodic shear layer was simulated on a low-resolution grid ($H = 128\Delta x$). Reprinted from Fig. 1 in [142] with permission from AIP Publishing.

spurious secondary vortices appear around $(x/H, y/H) = (0.75, 0.25)$ and $(0.25, 0.75)$. Therefore, the LBM obtains a non-physical result when simulating high Reynolds number flows on a low-resolution grid. In Fig. 2.6(b), no spurious secondary vortices appear but the vorticity contours are excessively dissipative. This occurs because hyper-viscosity in the LKS causes non-negligible dissipation error. In Fig. 2.6(c), no secondary vortices appear and the contours are sharp, indicating that the LWACM with $c_s = 0.7/\sqrt{3}$ suppresses the spurious vortices and also generates small dissipation error. However, similarly to the previous problem, the LWACM computation blows up when the speed of sound approaches that in the LBM, namely, $c_s = 1/\sqrt{3}$. In contrast, the improved LKS with $n = 1$ and $A_p = 0$ obtains a good result even when $c_s = 1/\sqrt{3}$ (Fig. 2.6(d)). Therefore, the improved

LKS achieves higher numerical stability and lower dissipation error than the other methods. Note also that as the number of iterations increases, the results of the improved LKS are the same as those of $n = 1$.

On the high-resolution grid ($H = 256\Delta x$), the LBM provides sharp vorticity contours without spurious secondary vortices, but the LKS yields unclear vorticity contours, indicating that even the high-resolution grid is insufficient for this method. Meanwhile, the LWACM and improved LKS achieve the same sharp vorticity contours without developing spurious secondary vortices. This result is expected, as both methods performed well on the low-resolution grid.

2.8 Comparison between Improved LKS and MRT-LBM

As demonstrated in Sec. 2.7, the improved LKS can compute high Reynolds number flows on low-resolution grids. In recent years, such problems have been solved by the MRT-LBM (see Appendix C) with the MRT model for collision terms. Improved MRT-based methods such as the cascaded collision model [38], the cumulant collision model [40], and the non-orthogonal MRT model [30] are also gaining attention. However, the MRT-LBM and its improved versions are spatially accurate only to second order, as in the improved LKS. Because the improved LKS is a fast and simple algorithm, it is considered as the better computational scheme. In fact, the present authors have successfully applied the improved LKS in two-phase flow simulations with high density ratios, as will be described in Chap. 4. We expect that the improved LKS will be widely used in future simulations of high Reynolds number flows.

Chapter 3

Immersed Boundary–Lattice Boltzmann Method (IB-LBM)

When an arbitrarily shaped object moves freely in a fluid, the flow is called a moving boundary flow. Moving boundary flows are among the most important subjects in fluid mechanics from both practical and theoretical perspectives. In the past, moving boundary flows were often simulated on a body-fitted grid or an unstructured grid, which must be reconstructed around the moving object at each time step. In general, regenerating the grid is computationally time-intensive.

In recent years, attention has returned to the immersed boundary method (hereafter referred to as the IBM) proposed by Peskin [106, 107] in the 1970s for moving boundary flows. The IBM handles the boundary condition on a moving object in a stationary structured grid. As explained in the previous chapters, the LBM and LKS are implemented on a square (two-dimensional) or cubic (three-dimensional) lattice, so both methods are compatible with the IBM.

This chapter begins with LBM-based methods for moving boundary flows, and then describes the immersed boundary–lattice Boltzmann method combining the LBM and IBM (hereafter referred to as the IB-LBM), which has often been used in recent years.

3.1 LBM-based Methods for Moving Boundary Flows

The application of LBM to moving boundary flows began in the 1990s, when the LBM was itself proposed. When handling moving boundary flows, the following points are especially important in the LBM and LKS, which are performed on a square or cubic lattice:

(P1) How can we satisfy the no-slip boundary condition?
(P2) How can we calculate the fluid force acting on the boundary?

Solving these problems on a regular lattice has concerned many researchers.

Since the pioneering work of Ladd [85, 86], the LBM has been applied to various calculations of moving boundary flows, mainly in particle transport and sedimentation. In this study, (P1) was resolved by a staircase-like boundary approximation and application of the bounce-back condition, whereas (P2) was resolved by calculating the momentum loss of the fictitious particles when bouncing back from the boundary (the so-called momentum exchange).

Inamuro et al. [62] resolved (P1) by applying the counter-slip condition to the boundary, which was also approximated by a staircase-like discretization (Sec. 1.11), and resolved (P2) by calculating the stress tensor from the velocity distribution function (see Eq. (1.50)). However, these studies require a staircase-like re-approximation of the boundary at each movement of the object, which is computationally intensive and lowers the numerical accuracy. Moreover, the boundary is impacted by non-physical oscillations in the fluid force whenever it crosses the lattice.

The improved bounce-back condition has improved the accuracy of (P1) computations. The improved bounce-back condition was first applied to moving boundary flows by Lallemand and Luo [89] as a solution to (P1). These authors resolved (P2) by considering the momentum exchange. Although the improved bounce-back condition has become a popular solution to moving boundary flows, the intersection of the boundary and the lattice must be determined at each movement of the object, which complicates the algorithm and counteracts its excellent numerical accuracy. Moreover, as occurs when discretizing the boundary curve with the staircase approximation, non-physical oscillations appear in the fluid force acting on the boundary.

The IB-LBM was first applied to moving boundary flows by Feng and Michaelides [32]. To solve (P1), they imposed no-slip conditions at the boundary points by applying an appropriate volume force near the boundary. That is, the movement of the boundary can be expressed by "movement of the volume force field." To solve (P2), they considered that the fluid force acting on the boundary equals the reaction of the volume force. This novel idea simplifies the calculation of moving boundary flows satisfying both (P1) and (P2). In addition, as the fluid force acting on the boundary is not subjected to non-physical oscillations, this idea overcomes the drawback of the previous methods. For these reasons, IB-LBMs have been applied in many simulations of moving boundary flows.

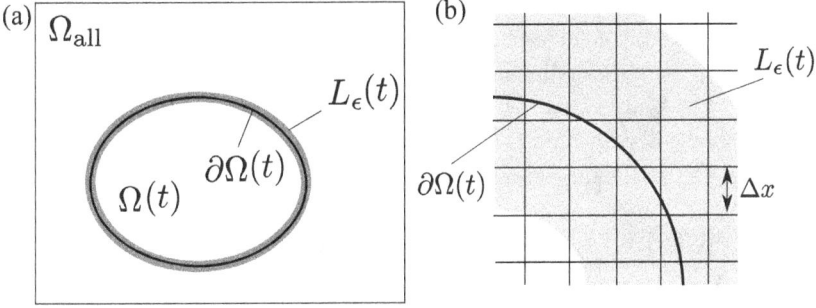

Fig. 3.1 (a) IBM representation of a moving object in a fluid [133]; (b) images of the object boundary $\partial\Omega(t)$ and boundary neighborhood region $L_\epsilon(t)$ on a square lattice.

3.2 Idea of IBM

As mentioned above, the boundary in IBM is represented as a "field" of volume force. This section describes the IBM in detail.

When an object $\Omega(t)$ exists in a region Ω_{all}, the IBM assumes that incompressible viscous fluid fills both the outside and inside of the object boundary $\partial\Omega(t)$ (Fig. 3.1(a)). To satisfy the no-slip condition on the object boundary, an appropriate volume force $\boldsymbol{G}(\boldsymbol{x}, t)$ is applied to the neighborhood $L_\epsilon(t)$ of the object boundary.

The boundary neighborhood region $L_\epsilon(t)$ is often set very close to the object boundary (within Δx–$2\Delta x$ on one side). When combined with the LBM or LKS on a square or cubic lattice, the no-slip condition on $\partial\Omega(t)$ can be satisfied by applying a volume force to the grid points in $L_\epsilon(t)$, as shown in Fig. 3.1(b). A moving object boundary is then easily represented by moving the "field" of the volume force. Meanwhile, the fluid force acting on the boundary can be calculated from the volume force field, as explained in Sec. 3.5. Note that the IBM can represent a complex moving boundary on a stationary structured grid, which is a distinct advantage of the method.

3.3 LBM with an External Forcing Term (Fractional-Step Method)

As mentioned above, the IBM satisfies the no-slip boundary condition on a moving object by applying a volume force determined by the flow velocity. However, as the volume force affects the flow velocity, both the volume force and flow velocity must be calculated in a coupled manner. As the coupling method, we here adopt the fractional-step method, which divides Eq. (1.51) into two steps. The first step evolves a temporary velocity distribution function without considering the external force. In the second step, this temporary velocity distribution function is corrected by an external force. Applying the volume force $G(x,t)$ as the external force,[1] the time evolution of the velocity distribution function $f_i(x,t)$ is computed as follows. In the following, we use the incompressible equilibrium distribution function $f_i^{eq,in}$.

Step 1. Evolve $f_i(x,t)$ in time without the external force:

$$f_i^*(x + c_i \Delta x, t + \Delta t) = f_i(x,t) - \frac{1}{\tau}\left[f_i(x,t) - f_i^{eq,in}(x,t)\right]. \quad (3.1)$$

Step 2. Correct f_i^* by the volume force G:

$$f_i(x, t + \Delta t) = f_i^*(x, t + \Delta t) + 3\Delta x E_i c_i \cdot G(x, t + \Delta t). \quad (3.2)$$

As mentioned in Sec. 1.6, we can use the above scheme to compute the Navier–Stokes equations with the external force term. The relative error of this computation is $O((\Delta x)^2)$. Although several other methods can account for the external force term in the IB-LBM (see [80]), the method presented here is the simplest and its numerical accuracy is comparable with those of more complex methods.

3.4 Formulation of the IBM

As mentioned in Sec. 3.2, the incompressible viscous fluid in the IBM fills both the outside and inside of the object, and the no-slip boundary condition is satisfied by applying an appropriate volume force on the lattice points near the object boundary (Fig. 3.1). The method that determines the volume force is the key part of the IBM, and various types of IBMs determine the volume force by different approaches [99]. In the following, we describe an IBM called the multi-direct forcing method (MDFM), which was proposed by Wang et al. [151]. The MDFM strongly enforces the no-slip condition on the object boundary by iteratively calculating the volume force.

---

[1]In Chap. 3, we suppose that the volume force $G(x,t)$ is the external force acting per unit volume. By using the fluid density as the reference density in the definition of dimensionless variables, this formulation is apparently that of the external force in the previous chapters.

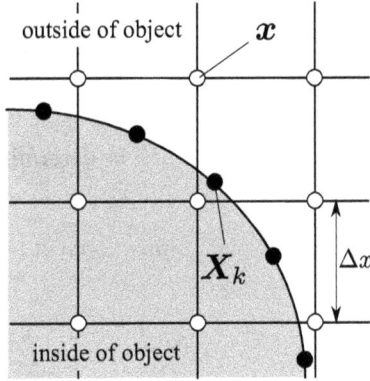

Fig. 3.2 Positional relationship between the boundary and lattice points in the IB-LBM. The boundary points X_k are defined independently of the lattice points x.

If $f_i(x,t)$, $u(x,t)$, and $p(x,t)$ are known, the temporary velocity distribution function $f_i^*(x,t+\Delta t)$ and the temporary flow velocity $u^*(x,t+\Delta t)$ can be computed by Eqs. (3.1) and (1.60), respectively. Now, let the boundary points that move with the object boundary and their velocities be $X_k(t+\Delta t)$ and $U_k(t+\Delta t)$ $(k = 1,2,\cdots,N_b)$, respectively. Here, N_b is the number of boundary points. In general, the boundary point X_k does not coincide with the lattice point x (see Fig. 3.2), so the temporary flow velocity $u^*(X_k,t+\Delta t)$ at the boundary point is interpolated from the flow velocities at the surrounding lattice points as follows:

$$u^*(X_k,t+\Delta t) = \sum_{x} u^*(x,t+\Delta t)\, W(x-X_k)\,(\Delta x)^d, \qquad (3.3)$$

where $\sum_{x}$ denotes summation over all lattice points x, and d represents the dimension ($d = 2$ and 3 in the two- and three-dimensional cases, respectively). The weighting function W is that proposed by Peskin [108] and is calculated as

$$W(x,y,z) = \frac{1}{\Delta x}w\left(\frac{x}{\Delta x}\right)\cdot\frac{1}{\Delta x}w\left(\frac{y}{\Delta x}\right)\cdot\frac{1}{\Delta x}w\left(\frac{z}{\Delta x}\right), \qquad (3.4)$$

where

$$w(r) = \begin{cases} \frac{1}{8}\left(3 - 2|r| + \sqrt{1 + 4|r| - 4r^2}\right), & |r| \le 1, \\ \frac{1}{8}\left(5 - 2|r| - \sqrt{-7 + 12|r| - 4r^2}\right), & 1 \le |r| \le 2, \\ 0, & \text{otherwise.} \end{cases} \qquad (3.5)$$

Note that the weight function in three-dimensional space is the product of three weight functions in one-dimensional space. In two-dimensional space, the weight function is obtained by removing $\frac{1}{\Delta x}w\left(\frac{z}{\Delta x}\right)$ from the three-dimensional weight function.

Next, the volume force $\boldsymbol{G}(\boldsymbol{x}, t + \Delta t)$ is determined by the following iterative computation.

Step 0. Compute the initial volume force at the boundary points of the object:

$$\boldsymbol{G}_0(\boldsymbol{X}_k, t + \Delta t) = \mathrm{Sh}\frac{U_k - \boldsymbol{u}^*(\boldsymbol{X}_k, t + \Delta t)}{\Delta t}, \tag{3.6}$$

where $\mathrm{Sh}/\Delta t = 1/\Delta x$ (see Appendix A).

Step 1. Compute the volume force at the lattice points during the ℓth iteration:

$$\boldsymbol{G}_\ell(\boldsymbol{x}, t + \Delta t) = \sum_{k=1}^{N_\mathrm{b}} \boldsymbol{G}_\ell(\boldsymbol{X}_k, t + \Delta t) \, W(\boldsymbol{x} - \boldsymbol{X}_k) \, \Delta V, \tag{3.7}$$

where ΔV is a small volume in which the volume force is applied, determined as $\Delta V = S/N_\mathrm{b} \times \Delta x$. Here, S is the surface area of the object and S/N_b is approximately equal to $(\Delta x)^{d-1}$.

Step 2. Correct the flow velocity at the lattice points:

$$\boldsymbol{u}_\ell(\boldsymbol{x}, t + \Delta t) = \boldsymbol{u}^*(\boldsymbol{x}, t + \Delta t) + \frac{\Delta t}{\mathrm{Sh}}\boldsymbol{G}_\ell(\boldsymbol{x}, t + \Delta t). \tag{3.8}$$

Step 3. Interpolate the flow velocity at the boundary points of the object:

$$\boldsymbol{u}_\ell(\boldsymbol{X}_k, t + \Delta t) = \sum_{\boldsymbol{x}} \boldsymbol{u}_\ell(\boldsymbol{x}, t + \Delta t) \, W(\boldsymbol{x} - \boldsymbol{X}_k) \, (\Delta x)^d. \tag{3.9}$$

Step 4. If the error in $\boldsymbol{u}_\ell(\boldsymbol{X}_k, t + \Delta t)$ from the no-slip boundary condition is large, then correct the volume force at the boundary points of the object:

$$\boldsymbol{G}_{\ell+1}(\boldsymbol{X}_k, t + \Delta t) = \boldsymbol{G}_\ell(\boldsymbol{X}_k, t + \Delta t) + \mathrm{Sh}\frac{U_k - \boldsymbol{u}_\ell(\boldsymbol{X}_k, t + \Delta t)}{\Delta t}. \tag{3.10}$$

Return to **Step 1**.

If the error from the no-slip boundary condition is sufficiently small, then go to **Step 5**.

Step 5. Determine the volume force at the lattice points at time $(t + \Delta t)$:

$$G(x, t + \Delta t) = G_\ell(x, t + \Delta t). \tag{3.11}$$

Preliminary calculations showed that $G_{\ell=5}(x, t + \Delta t)$ sufficiently satisfies the no-slip condition at the boundary points of the object [133]. Therefore, in the following numerical examples, we set $\ell = 5$. However, when the relaxation time $\tau > 1$, a large flow velocity slip occurs at the boundary [121], which exerts a large and persistent effect even after increasing the number of iterations of the volume force computation. In practical problems, this situation rarely arises because τ is usually less than 1.

3.5 Force and Torque Acting on a Moving Object

Regarding the volume force $G(x, t)$ computed in the previous section as the force applied to the fluid around an object, we can calculate the force acting on the object from the law of action and reaction. For this purpose we define $F_{tot}(t)$, which is the negative of the summed volume forces $G(x, t)$ [87]:

$$F_{tot}(t) = -\sum_x G(x, t) \, (\Delta x)^d. \tag{3.12}$$

However, it should be noted that $F_{tot}(t)$ is the force received from the fluids both inside and outside the object boundary. To calculate the force $F(t)$ acting on the object, we should subtract the force received from the internal fluid from $F_{tot}(t)$ [146, 149]. If the volume force applied to the internal fluid is $F_{in}(t)$, the force received from the internal fluid is $-F_{in}(t)$ (by the law of action and reaction). Since $F_{in}(t)$ is equal to the time derivative of the linear momentum of the internal fluid, it can be calculated as follows:

$$F_{in}(t) = \frac{d}{dt} \int_{x \in \Omega(t)} u(x, t) dx. \tag{3.13}$$

The specific calculation will be described later. Therefore, the force $F(t)$ acting on the object is given by

$$F(t) = F_{tot}(t) + F_{in}(t). \tag{3.14}$$

Similarly, the torque $T(t)$ acting on the object around its center of mass $X_c(t)$ is calculated by the following equations:

$$T(t) = T_{\text{tot}}(t) + T_{\text{in}}(t), \tag{3.15}$$

$$T_{\text{tot}}(t) = -\sum_{\boldsymbol{x}} [\boldsymbol{x} - \boldsymbol{X}_{\text{c}}(t)] \times \boldsymbol{G}(\boldsymbol{x}, t) \, (\Delta x)^d, \tag{3.16}$$

$$T_{\text{in}}(t) = \frac{\mathrm{d}}{\mathrm{d}t} \int_{\boldsymbol{x} \in \Omega(t)} [\boldsymbol{x} - \boldsymbol{X}_{\text{c}}(t)] \times \boldsymbol{u}(\boldsymbol{x}, t) \mathrm{d}\boldsymbol{x}. \tag{3.17}$$

From the above, we find that to calculate the force and torque acting on the moving object, we need to calculate $\boldsymbol{F}_{\text{in}}(t)$ and $\boldsymbol{T}_{\text{in}}(t)$. Hereafter, these quantities are collectively called the "internal mass effect." Among several methods for approximating the internal mass effect [33, 149], we select the most direct approximation, the Lagrangian point approximation [133]. Defining many Lagrangian points $\boldsymbol{X}_{\text{in}}(t)$ inside the object that move with the object, we calculate Eqs. (3.13) and (3.17) by adding the linear and angular momentums of the internal mass over these points. The number of Lagrangian points approximates the number of lattice points within the internal region of the object. As a Lagrangian point $\boldsymbol{X}_{\text{in}}(t)$ does not generally coincide with the lattice points, the velocity $\boldsymbol{u}(\boldsymbol{X}_{\text{in}}, t)$ at that point must be interpolated from the velocities at the surrounding lattice points:

$$\boldsymbol{u}(\boldsymbol{X}_{\text{in}}, t) = \sum_{\boldsymbol{x}} \boldsymbol{u}(\boldsymbol{x}, t) \, W(\boldsymbol{x} - \boldsymbol{X}_{\text{in}}(t)) \, (\Delta x)^d. \tag{3.18}$$

The linear and angular momentums $\boldsymbol{P}_{\text{in}}(t)$ and $\boldsymbol{L}_{\text{in}}(t)$ of the internal mass are calculated by

$$\boldsymbol{P}_{\text{in}}(t) = \sum_{\text{all} \boldsymbol{X}_{\text{in}}(t)} \boldsymbol{u}(\boldsymbol{X}_{\text{in}}, t) \, (\Delta x)^d, \tag{3.19}$$

$$\boldsymbol{L}_{\text{in}}(t) = \sum_{\text{all} \boldsymbol{X}_{\text{in}}(t)} [\boldsymbol{X}_{\text{in}}(t) - \boldsymbol{X}_{\text{c}}(t)] \times \boldsymbol{u}(\boldsymbol{X}_{\text{in}}, t) \, (\Delta x)^d. \tag{3.20}$$

The time derivatives in $\boldsymbol{F}_{\text{in}}(t)$ and $\boldsymbol{T}_{\text{in}}(t)$ are discretized using their values at times $(t - \Delta t)$ and t.

$$\boldsymbol{F}_{\text{in}}(t) \simeq \mathrm{Sh} \frac{\boldsymbol{P}_{\text{in}}(t) - \boldsymbol{P}_{\text{in}}(t - \Delta t)}{\Delta t}, \tag{3.21}$$

$$\boldsymbol{T}_{\text{in}}(t) \simeq \mathrm{Sh} \frac{\boldsymbol{L}_{\text{in}}(t) - \boldsymbol{L}_{\text{in}}(t - \Delta t)}{\Delta t}. \tag{3.22}$$

At the initial time $t = 0$, we assume that $\boldsymbol{P}_{\text{in}}(-\Delta t) = \boldsymbol{P}_{\text{in}}(0)$ and $\boldsymbol{L}_{\text{in}}(-\Delta t) = \boldsymbol{L}_{\text{in}}(0)$. Note that when the internal region of the object has no volume (e.g., the object is a very thin rod or a very thin board), the internal mass effect is absent, and $\boldsymbol{F}_{\text{in}} = \boldsymbol{T}_{\text{in}} = \boldsymbol{0}$.

3.6 Motion of the Boundary Lagrangian Points

If the object is freely moving, the position $X_k(t)$ and velocity $U_k(t)$ of a boundary point moving with the object are determined by solving the equation of motion of the object. In the following explanation, we describe a two-dimensional rigid body moving in a fluid by its position $X_c(t)$ and velocity $U_c(t)$ of its center of mass, and by the angle $\Theta_c(t)$ and angular velocity $\Omega_c(t)$ around the center of mass. If the object is a three-dimensional rigid body, we must apply a three-dimensional coordinate transformation using Euler angles or quaternions. The description is complicated so is omitted here. Interested readers are referred to [133, 136].

The position $X_c(t)$ and angle $\Theta_c(t)$ are respectively related to the center-of-mass velocity $U_c(t)$ and the angular velocity $\Omega_c(t)$ around the center of mass are given by

$$\text{Sh}\frac{\mathrm{d}X_c}{\mathrm{d}t} = U_c, \tag{3.23}$$

$$\text{Sh}\frac{\mathrm{d}\Theta_c}{\mathrm{d}t} = \Omega_c. \tag{3.24}$$

Because these equations are described on the diffusive time scale (as in the LBM), the time derivative terms are multiplied by the Strouhal number Sh. Invoking Newton's equations of motion, we also determine the velocity $U_c(t)$ and angular velocity $\Omega_c(t)$ from the force $F(t)$ and torque $T(t)$, respectively, as follows:

$$M\,\text{Sh}\frac{\mathrm{d}U_c}{\mathrm{d}t} = F, \tag{3.25}$$

$$I\,\text{Sh}\frac{\mathrm{d}\Omega_c}{\mathrm{d}t} = T_z, \tag{3.26}$$

where M and I are the mass and inertia moment of the object, respectively, and T_z is the z-component of the torque T.

The above equations of motion of the object (3.23)–(3.26) must be coupled to the equations of motion of the fluid. For this purpose, we use a weak-coupling scheme which alternately evolves each difference equation in time with the same time step. Therefore, the equations of motion of the object are explicitly evolved in time and discretized with the time step Δt used in the LBM. In the Euler scheme, the difference equations of motion of the object are given by

$$\boldsymbol{X}_{\mathrm{c}}(t + \Delta t) = \boldsymbol{X}_{\mathrm{c}}(t) + \frac{\Delta t}{\mathrm{Sh}} \boldsymbol{U}_{\mathrm{c}}(t), \tag{3.27}$$

$$\Theta_{\mathrm{c}}(t + \Delta t) = \Theta_{\mathrm{c}}(t) + \frac{\Delta t}{\mathrm{Sh}} \Omega_{\mathrm{c}}(t), \tag{3.28}$$

$$\boldsymbol{U}_{\mathrm{c}}(t + \Delta t) = \boldsymbol{U}_{\mathrm{c}}(t) + \frac{\Delta t}{\mathrm{Sh}} \boldsymbol{F}(t)/M, \tag{3.29}$$

$$\Omega_{\mathrm{c}}(t + \Delta t) = \Omega_{\mathrm{c}}(t) + \frac{\Delta t}{\mathrm{Sh}} T_z(t)/I. \tag{3.30}$$

To perform the computation to second-order accuracy in Δt, one could apply the second-order Adams–Bashforth method. However, as the LBM is only first-order accurate in Δt, computing the equation of motion of the object to higher-order accuracy is not necessary.

After updating the position and velocity of the center of mass, along with the angle and angular velocity as described above, we can update the position $\boldsymbol{X}_k$ and velocity $\boldsymbol{U}_k$ of the boundary Lagrangian points as follows:

$$\boldsymbol{X}_k(t + \Delta t) = \boldsymbol{X}_{\mathrm{c}}(t + \Delta t) + R(t + \Delta t) \left[\boldsymbol{X}_k(0) - \boldsymbol{X}_{\mathrm{c}}(0) \right], \tag{3.31}$$

$$\boldsymbol{U}_k(t + \Delta t) = \boldsymbol{U}_{\mathrm{c}}(t + \Delta t) + \mathrm{Sh} \frac{dR}{dt}(t + \Delta t) \left[\boldsymbol{X}_k(0) - \boldsymbol{X}_{\mathrm{c}}(0) \right], \tag{3.32}$$

where R is the two-dimensional rotational matrix given by

$$R(t) = \begin{bmatrix} \cos \Theta_{\mathrm{c}}(t) & -\sin \Theta_{\mathrm{c}}(t) \\ \sin \Theta_{\mathrm{c}}(t) & \cos \Theta_{\mathrm{c}}(t) \end{bmatrix}. \tag{3.33}$$

The time derivative dR/dt of this matrix is given by

$$\mathrm{Sh} \frac{dR}{dt}(t) = \Omega_{\mathrm{c}}(t) \begin{bmatrix} -\sin \Theta_{\mathrm{c}}(t) & -\cos \Theta_{\mathrm{c}}(t) \\ \cos \Theta_{\mathrm{c}}(t) & -\sin \Theta_{\mathrm{c}}(t) \end{bmatrix}. \tag{3.34}$$

When the motion of the object is given in advance, i.e., when $\boldsymbol{X}_{\mathrm{c}}(t)$, $\boldsymbol{U}_{\mathrm{c}}(t)$, $\Theta_{\mathrm{c}}(t)$, and $\Omega_{\mathrm{c}}(t)$ are given, we can omit solving the equations of motion of the object. Instead, we update the position and velocity of the boundary points by Eqs. (3.31) and (3.32), respectively.

3.7　Computational Algorithm of IB-LBM

The computational algorithm of the IB-LBM is summarized below.

Step 0. Give the initial value of $f_i(\boldsymbol{x}, 0)$, and compute $p(\boldsymbol{x}, 0)$ and $\boldsymbol{u}(\boldsymbol{x}, 0)$ by Eqs. (1.59) and (1.60), respectively.

Step 1. Compute the equations of motion of the object (3.27)–(3.30), and compute $X_k(t + \Delta t)$ and $U_k(t + \Delta t)$ by Eqs. (3.31) and (3.32), respectively.

Step 2. Compute $f_i^*(x, t + \Delta t)$ and $u^*(x, t + \Delta t)$ by Eqs. (3.1) and (1.60), respectively. Then compute $u^*(X_k, t + \Delta t)$ by Eq. (3.3).

Step 3. Iteratively compute $G(x, t + \Delta t)$ by Eqs. (3.6)–(3.11).

Step 4. Compute $f_i(x, t + \Delta t)$ by Eq. (3.2), and compute $p(x, t + \Delta t)$ and $u(x, t + \Delta t)$ by Eqs. (1.59) and (1.60), respectively.

Step 5. Advance the time by Δt and return to **Step 1**.

Recall from Chap. 1 that the LBM is second-order accurate in space. However, the IB-LBM has only first-order accuracy in space because the velocity gradient is discontinuous across the boundary. A higher-order IB-LBM has been proposed by Suzuki and Inamuro [134], but the computational algorithm is rather complicated.

3.8 Thermal Immersed Boundary–Lattice Boltzmann Method (Thermal IB-LBM)

The above concepts developed for the IB-LBM in flow fields can be extended to thermal fields. That is, we replace the boundary condition in Fig. 3.1 with a thermal boundary condition on $\partial\Omega(t)$. To this end, we generate an appropriate amount of heat $Q(x, t)$ per unit volume in the boundary neighborhood region $L_\epsilon(t)$.

Identically to the IB-LBM for the flow field, we compute the velocity distribution function $g_i(x, t)$ of the temperature field by the fractional-step method. If the amount of heat generated per unit volume is $Q(x, t)$, then $g_i(x, t)$ evolves in time as follows.

Step 1. Evolve $g_i(x, t)$ in time without heat generation:

$$g_i^*(x + c_i \Delta x, t + \Delta t) = g_i(x, t) - \frac{1}{\tau_g}\left[g_i(x, t) - g_i^{\text{eq}}(x, t)\right]. \quad (3.35)$$

Step 2. Correct g_i^* by the heat-generation term Q:

$$g_i(x, t + \Delta t) = g_i^*(x, t + \Delta t) + \Delta x \frac{\alpha}{k} E_i Q(x, t + \Delta t), \quad (3.36)$$

where α and k are the temperature and thermal conductivity coefficients, respectively. The quantity k/α defines the apparent heat capacity.[2]

The thermal boundary conditions are classified into two types, i.e., the isothermal and isoheat-flux conditions (called the Dirichlet and Neumann conditions, respectively). In the following, we calculate the amount of heat $Q(\boldsymbol{x},t)$ generated per unit volume (hereafter referred to as the heat-source term) for each boundary condition. We introduce the methods proposed by Wang et al. [152] and Ren et al. [112] for the isothermal and isoheat-flux conditions, respectively. The boundary is represented by discrete boundary points (as in the IBM for the flow field (see Fig. 3.2)).

(1) Isothermal condition

This method is based on the MDFM developed for the IBM for the flow field. By iteratively calculating the heat-source term, it strongly enforces the isothermal condition on the object boundary.

When $g_i(\boldsymbol{x},t)$, $\boldsymbol{u}(\boldsymbol{x},t)$, and $T(\boldsymbol{x},t)$ are known, the temporary velocity distribution function $g_i^*(\boldsymbol{x},t+\Delta t)$ and temporary temperature $T^*(\boldsymbol{x},t+\Delta t)$ can be computed by Eqs. (3.35) and (1.62), respectively. The temporary temperature $T^*(\boldsymbol{X}_k,t+\Delta t)$ at the boundary point is interpolated from the temperatures at the surrounding lattice points as follows:

$$T^*(\boldsymbol{X}_k,t+\Delta t) = \sum_{\boldsymbol{x}} T^*(\boldsymbol{x},t+\Delta t)\, W(\boldsymbol{x}-\boldsymbol{X}_k)\,(\Delta x)^d, \qquad (3.37)$$

where the weighting function W is given by Eq. (3.4) in the IBM for the flow field.

Let $T_k^{\mathrm{d}}(t+\Delta t)$ be the desired temperature at boundary point $\boldsymbol{X}_k$. The heat-source term $Q(\boldsymbol{x},t+\Delta t)$ is iteratively computed by the following procedure:

Step 0. Compute the initial heat-source terms at the boundary points of the object:

$$Q_0(\boldsymbol{X}_k,t+\Delta t) = \frac{k}{\alpha}\mathrm{Sh}\frac{T_k^{\mathrm{d}} - T^*(\boldsymbol{X}_k,t+\Delta t)}{\Delta t}, \qquad (3.38)$$

where $\mathrm{Sh}/\Delta t = 1/\Delta x$ (see Appendix A).

---

[2]From the definition of the dimensionless variables in Appendix A, we originally have $k/\alpha = 1$, but from Eqs. (1.67) and (1.70), we have $k/\alpha \neq 1$ in the LBM, indicating an apparent heat capacity.

Step 1. Compute the heat-source terms at the lattice points during the mth iteration:

$$Q_m(x, t + \Delta t) = \sum_{k=1}^{N_b} Q_m(X_k, t + \Delta t) \, W(x - X_k) \, \Delta V, \quad (3.39)$$

where N_b is the number of boundary points. Identically to the IBM for the flow field, we set $\Delta V = S/N_b \times \Delta x$ and $S/N_b \simeq (\Delta x)^{d-1}$.

Step 2. Correct the temperature at the lattice points:

$$T_m(x, t + \Delta t) = T^*(x, t + \Delta t) + \frac{\alpha}{k} \frac{\Delta t}{\mathrm{Sh}} Q_m(x, t + \Delta t). \quad (3.40)$$

Step 3. Interpolate the temperature at the boundary points of the object:

$$T_m(X_k, t + \Delta t) = \sum_x T_m(x, t + \Delta t) \, W(x - X_k) \, (\Delta x)^d. \quad (3.41)$$

Step 4. If the error in $T_m(X_k, t + \Delta t)$ from the isothermal boundary condition is large, then correct the heat-source term at the boundary points of the object:

$$\begin{aligned} Q_{m+1}(X_k, t + \Delta t) &= Q_m(X_k, t + \Delta t) \\ &+ \frac{k}{\alpha} \mathrm{Sh} \frac{T_k^d - T_m(X_k, t + \Delta t)}{\Delta t}, \end{aligned} \quad (3.42)$$

and return to **Step 1**.

If the error from the isothermal boundary condition is sufficiently small, then go to **Step 5**.

Step 5. Determine the heat-source term at the lattice points at time $(t + \Delta t)$:

$$Q(x, t + \Delta t) = Q_m(x, t + \Delta t). \quad (3.43)$$

Preliminary calculations showed that $Q_{m=5}(x, t + \Delta t)$ sufficiently satisfies the isothermal condition at the boundary points of the object [138]. Therefore, in the following numerical examples, we set $m = 5$. However, at relaxation times τ_g exceeding 1, a large temperature jump occurs at the boundary [120]. The effect of this temperature jump is large and persists even after increasing the number of iterations for determining the heat-source term. In typical problems, τ_g is less than 1 so this situation does not arise.

(2) Isoheat-flux condition

When $g_i(\boldsymbol{x}, t)$, $\boldsymbol{u}(\boldsymbol{x}, t)$, and $T(\boldsymbol{x}, t)$ are known, the temporary velocity distribution function $g_i^*(\boldsymbol{x}, t + \Delta t)$ can be computed by Eq. (3.35), and the temporary heat-flux vector $\boldsymbol{q}^*(\boldsymbol{x}, t + \Delta t)$ can be computed by Eq. (1.68) as follows:

$$\boldsymbol{q}^*(\boldsymbol{x}, t + \Delta t) = \sum_{i=1}^{N} g_i^*(\boldsymbol{x}, t + \Delta t) \left[\boldsymbol{c}_i - \boldsymbol{u}(\boldsymbol{x}, t + \Delta t) \right]. \tag{3.44}$$

The temporary heat-flux vector $\boldsymbol{q}^*(\boldsymbol{X}_k, t + \Delta t)$ at a boundary point is interpolated from the heat-flux vectors at the surrounding lattice points:

$$\boldsymbol{q}^*(\boldsymbol{X}_k, t + \Delta t) = \sum_{\boldsymbol{x}} \boldsymbol{q}^*(\boldsymbol{x}, t + \Delta t) \, W(\boldsymbol{x} - \boldsymbol{X}_k) \, (\Delta x)^d. \tag{3.45}$$

Let $\boldsymbol{n}_k(t + \Delta t)$ be the unit normal vector to the boundary at boundary point $\boldsymbol{X}_k(t + \Delta t)$ pointing toward the external fluid (Fig. 3.3). The temporary heat flux in the normal direction is then given by

$$q_{\mathrm{n}}^*(\boldsymbol{X}_k, t + \Delta t) = \boldsymbol{n}_k \cdot \boldsymbol{q}^*(\boldsymbol{X}_k, t + \Delta t). \tag{3.46}$$

In the original method proposed by Ren et al. [112], the temperature gradient in the temporary heat flux is calculated by a second-order central difference approximation. In the present method, it is calculated by Eq. (3.44) using the velocity distribution functions. In our preliminary two-dimensional calculations, both methods gave almost the same result [138].

Let $q_k^{\mathrm{d}}(t + \Delta t)$ be the desired heat flux in the normal direction at boundary point $\boldsymbol{X}_k$ (see Fig. 3.3). The heat-source term $Q(\boldsymbol{x}, t + \Delta t)$ is determined by

$$Q(\boldsymbol{x}, t + \Delta t) = \sum_{k=1}^{N_{\mathrm{b}}} Q(\boldsymbol{X}_k, t + \Delta t) \, W(\boldsymbol{x} - \boldsymbol{X}_k) \, \Delta V, \tag{3.47}$$

where

$$Q(\boldsymbol{X}_k, t + \Delta t) = 2 \frac{q_k^{\mathrm{d}}(t + \Delta t) - q_{\mathrm{n}}^*(\boldsymbol{X}_k, t + \Delta t)}{\Delta x}. \tag{3.48}$$

The factor 2 in the right-hand side of the above equation accounts for the dual effect (inside and outside the object) of the heat flux induced by the difference between $q_k^{\mathrm{d}}(t + \Delta t)$ and $q_{\mathrm{n}}^*(\boldsymbol{X}_k, t + \Delta t)$.

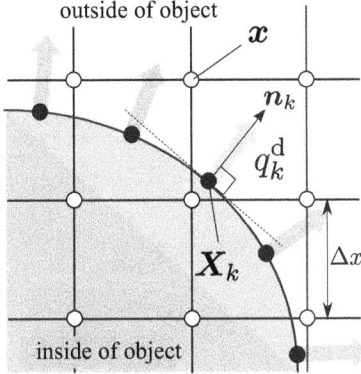

Fig. 3.3 Thermal IB-LBM under the isoheat-flux condition. The boundary Lagrangian points X_k are defined independently of the lattice points x. At each boundary point, we define the unit normal vector n_k and specify the desired heat flux q_k^d.

3.9 Multi-block Grid Method

The multi-block grid method is popular because it saves computational time while maintaining the numerical accuracy. This method applies a fine grid near the moving boundary where the flow velocity and pressure changes are large, and a coarse grid far from the boundary. In this section, we apply the multi-block grid method [73] to the moving boundary problem.

Figure 3.4 shows a multi-block grid on which a fine grid (of width Δx_f)

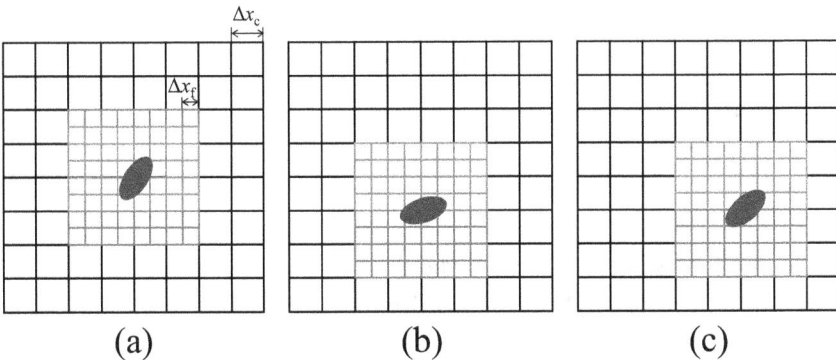

Fig. 3.4 Example of a multi-block grid combining a moving fine grid Δx_f with a stationary coarse grid $\Delta x_c (= 2\Delta x_f)$. Note that both grids overlap by Δx_c. When an object moves by more than Δx_c in the x- or y-direction, the fine grid is translated by Δx_c in that direction.

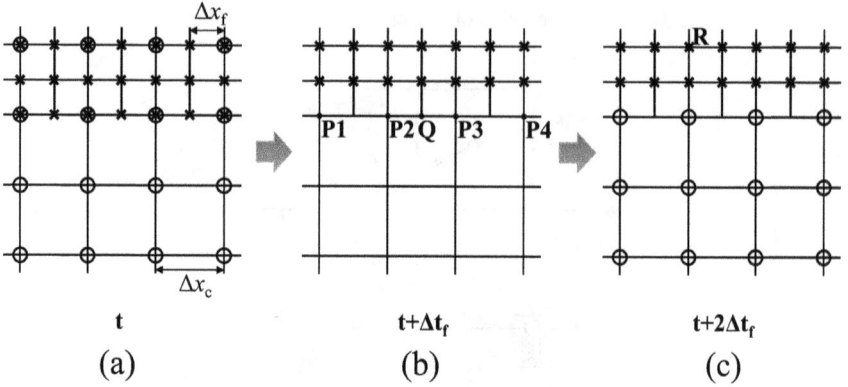

Fig. 3.5   Passing data in the multi-block grid. The symbols ∘ and × represent the coarse and fine grid points, respectively, on which the velocity distribution functions are obtained.

moves with the moving object while a coarse grid (of width $\Delta x_c (= 2\Delta x_f)$) is fixed in space. Note that the grids overlap exactly on a width of Δx_c. Furthermore, the fine grid is translated by Δx_c when the object moves by more than Δx_c in the x- or y-direction. In Fig. 3.4, the fine grid shifts downwards from (a) to (b) by Δx_c and rightwards from (b) to (c) by Δx_c.

To match the calculated results on both grids, we must first match the fictitious particle velocities on both grids. For this purpose, we define the following relationship between the grid width Δx and time step Δt:

$$\frac{\Delta x_c}{\Delta t_c} = \frac{\Delta x_f}{\Delta t_f}. \tag{3.49}$$

We must then match the viscosity coefficients on both grids, which requires the following relationship:

$$\left(\tau_c - \frac{1}{2}\right)\Delta x_c = \left(\tau_f - \frac{1}{2}\right)\Delta x_f. \tag{3.50}$$

Finally, we must fit the velocity distribution functions in the area overlapped by both grids. To ensure continuity of the flow velocity, pressure, and stress tensor in the overlapping grid points, we specify that the velocity distribution function f_i^c on the coarse grid equals the velocity distribution function f_i^f on the fine grid [35]:

$$f_i^{eq,c} = f_i^{eq,f}, \tag{3.51}$$

$$f_i^c = \frac{\tau_c}{\tau_f} \frac{\Delta x_c}{\Delta x_f} f_i^f + \left(1 - \frac{\tau_c}{\tau_f} \frac{\Delta x_c}{\Delta x_f}\right) f_i^{\mathrm{eq,c}}. \tag{3.52}$$

As the LBM is a hyperbolic-type computational scheme, we cannot compute the inward-facing velocity distribution function at the outmost grid points on each grid at the next time step (see Fig. 3.5). Therefore, at the next time step, we instead estimate the unknown velocity distribution function at the outmost grid points on one grid from the known velocity distribution function at the outmost grid points on the other grid. That is, at time $(t + \Delta t_f)$ in Fig. 3.5(b), we estimate $f_i^f(\boldsymbol{x}_{\mathrm{P}k}, t + \Delta t_f)$ using $[f_i^c(\boldsymbol{x}_{\mathrm{P}k}, t) + f_i^c(\boldsymbol{x}_{\mathrm{P}k}, t + \Delta t_c)]/2$ and Eq. (3.52). Next, using the four points $f_i^f(\boldsymbol{x}_{\mathrm{P}k}, t + \Delta t_f)$ around Point Q, we interpolate $f_i^f(\boldsymbol{x}_{\mathrm{Q}}, t + \Delta t_f)$ with numerical errors $O((\Delta x)^4)$. Similarly, at time $(t + 2\Delta t_f)$ in Fig. 3.5(c), we estimate $f_i^f(\boldsymbol{x}_{\mathrm{P}k}, t + 2\Delta t_f)$ using $f_i^c(\boldsymbol{x}_{\mathrm{P}k}, t + \Delta t_c)$ and Eq. (3.52), and interpolate $f_i^f(\boldsymbol{x}_{\mathrm{Q}}, t + 2\Delta t_f)$ using $f_i^f(\boldsymbol{x}_{\mathrm{P}k}, t + 2\Delta t_f)$. In addition, we estimate $f_i^c(\boldsymbol{x}_{\mathrm{R}}, t + \Delta t_c)$ using $f_i^f(\boldsymbol{x}_{\mathrm{R}}, t + 2\Delta t_f)$ and Eq. (3.52).

The procedure of the multi-block grid method is summarized below.

Step 0. Give the initial velocity distribution functions $f_i^f(\boldsymbol{x}, 0)$ and $f_i^c(\boldsymbol{x}, 0)$ at the grid points of both grids. At this time, Eq. (3.52) should be satisfied where the grid points overlap.

Step 1. Compute $f_i^f(\boldsymbol{x}, t + \Delta t_f)$ and $f_i^c(\boldsymbol{x}, t + \Delta t_c)$ using the computational algorithm of the LBM.

Step 2. Estimate $f_i^f(\boldsymbol{x}_{\mathrm{P}k}, t + \Delta t_f)$ and interpolate $f_i^f(\boldsymbol{x}_{\mathrm{Q}}, t + \Delta t_f)$.

Step 3. Compute $f_i^f(\boldsymbol{x}, t + 2\Delta t_f)$ using the computational algorithm of the LBM.

Step 4. Estimate $f_i^f(\boldsymbol{x}_{\mathrm{P}k}, t + 2\Delta t_f)$ and $f_i^c(\boldsymbol{x}_{\mathrm{R}}, t + \Delta t_c)$, and interpolate $f_i^f(\boldsymbol{x}_{\mathrm{Q}}, t + 2\Delta t_f)$.

Step 5. Return to **Step 1**.

As described above, the fine grid is translated by Δx_c when the object moves by more than Δx_c in the x- or y-direction. When moving the fine grid, we generate new fine grid points ahead of the moving direction. The velocity distribution functions of these new grid points are estimated by interpolating between the velocity distribution functions of the coarse grid points as explained above.

The multi-block grid method is applicable not only to moving objects in the IB-LBM, but also to calculations of flows around a stationary object and to calculations of bubbles or droplets in a two-phase LBM as described later.

3.10   Numerical Examples

3.10.1   *Sedimentation of an elliptical cylinder*

As a basic example of moving boundary flows that can be simulated in the IB-LBM, we consider the sedimentation of an elliptical cylinder. This problem was investigated in detail by Xia et al. [155], who applied an LBM with the improved bounce-back condition. An elliptical cylinder with major and minor axis lengths of a and b, respectively, settles under gravitational acceleration g in a stationary fluid. The fluid fills a channel between parallel plates of width H (see Fig. 3.6). The computational domain is divided into a square lattice with lattice spacing Δx, H is set to $200\Delta x$, and the channel length is $17.5H$. The bounce-back condition is imposed on all boundaries of the computational domain. The geometric parameters of the elliptical cylinder are set to $a = H/4$ and $b = a/2$. Let $\gamma = \rho_b/\rho_f$ be the density ratio of the elliptical cylinder to the fluid. The mass and inertia moment of the elliptical cylinder are given by $M = \rho_b(\pi ab/4)$ and $I = M(a^2 + b^2)/16$, respectively. The net gravitational force acting on the elliptical cylinder is $F_g = (1 - 1/\gamma)Mg$. The internal mass effect is computed using the Lagrangian points approximation introduced in Sec. 3.5. The fluid is initially at rest and the elliptical cylinder begins moving from its initial position $(x_c, y_c) = (0, 0.5H)$ at an initial angle of $\theta = 45°$. The governing parameters of the system are the density ratio $\gamma = \rho_b/\rho_f$ and the Reynolds number $\mathrm{Re} = u_t a/\nu$, where u_t is the terminal velocity of the elliptical cylinder in the vertical direction. In the following, we show the results of $\gamma = 1.5$ and $\mathrm{Re} = 32.9$.

Figure 3.7 shows the flow field around the falling elliptical cylinder. Note that during the fall, the elliptical cylinder oscillates in the horizontal direction while separated vortices develop behind it. These phenomena are representative of a moving object in a fluid (such as fluttering leaves), whereby the flow induced by the motion changes the motion of the object.

To confirm the magnitude of the internal mass effect, we neglect the internal mass in Calculation (A) and approximate the motion of the internal fluid by a rigid body motion as in Uhlmann [149] (Calculation (B-1)) and Feng and Michaelides [33] (Calculation (B-2)). The results are compared with those of a calculation using the Lagrangian point approximation (Calculation (C)). Figure 3.8 shows the trajectory $(x/H, y/H)$ of the center of the elliptical cylinder and the time variation of the angle θ. The results of Calculations (B-1), (B-2), and (C), which account for the internal mass

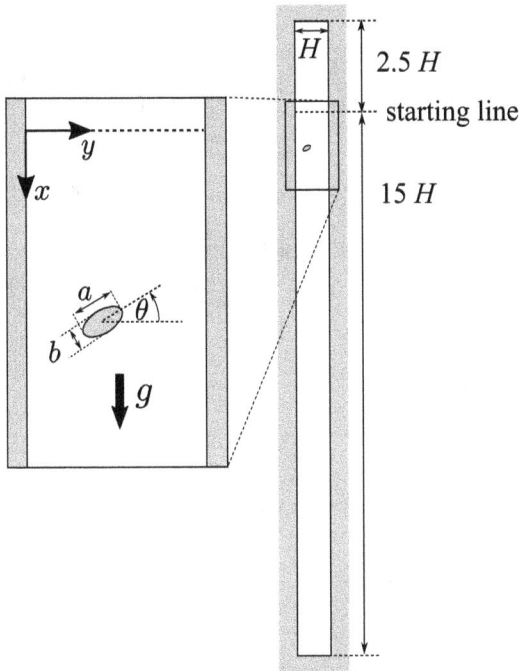

Fig. 3.6 Computational domain of an elliptical cylinder falling in a closed narrow domain. The elliptical cylinder is driven by a constant gravitational acceleration g. The x and y coordinates define the directions of gravity and channel width, respectively. The angle θ is the cylinder's angle of inclination with respect to the y-axis.

effect, are almost coincident. The result of Calculation (A), which neglects the internal mass effect, significantly differs from the other results in terms of both frequency and amplitude of the oscillation. Therefore, the internal mass exerts a significant effect in unsteady moving boundary flows. The coincidence of Calculations (B-1), (B-2), and (C) can be attributed to the low rotational Reynolds number (approximately 10, calculated as $\mathrm{Re_r} = a^2|\omega_{z,\mathrm{max}}|/(2\nu)$, where $|\omega_{z,\mathrm{max}}|$ is the magnitude of the maximum angular velocity of the elliptical cylinder) and the insignificant error in approximating the motion of the internal fluid by the rigid body motion.

3.10.2 *Sedimentation of a cold circular cylinder in a hot channel with natural convection*

To numerically demonstrate the thermal IB-LBM, we simulate the sedimentation of a cold circular cylinder in a hot channel. This problem has

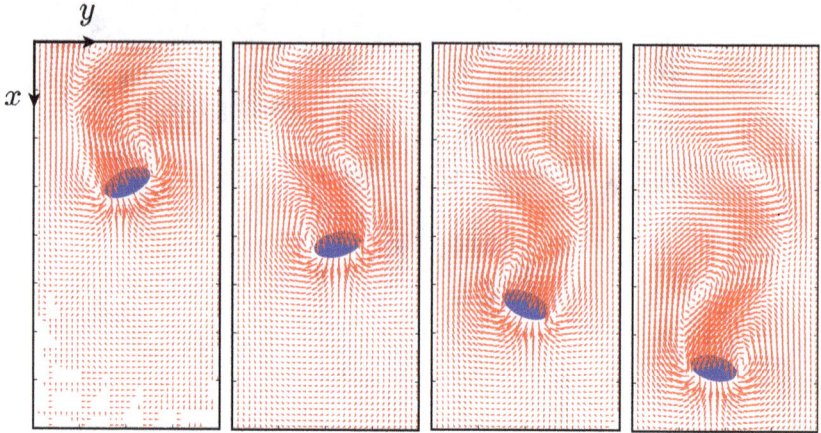

Fig. 3.7 Flow field around a falling elliptical cylinder (density ratio $\gamma = 1.5$, Reynolds number Re = 32.9) [133]. Reprinted from Fig. 9 in [133] with permission from Elsevier.

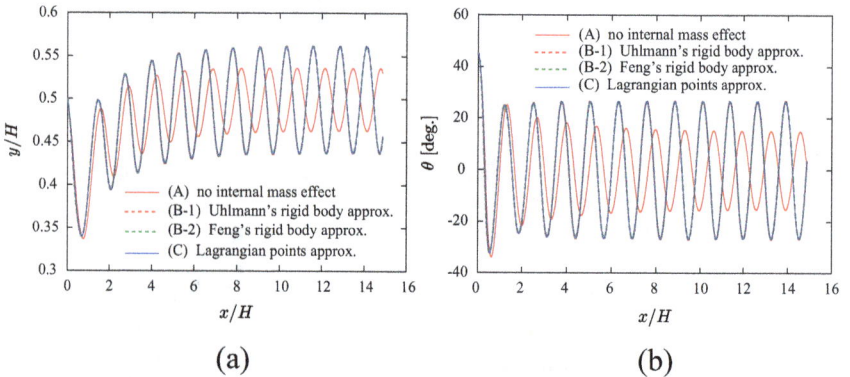

Fig. 3.8 (a) Trajectory of the center of an elliptical cylinder ($x/H, y/H$); (b) time variation of the angle θ (density ratio $\gamma = 1.5$, Reynolds number Re = 32.9) [133]. Reprinted from Fig. 10 in [133] with permission from Elsevier.

been widely used as a benchmark problem (for a recent example, see [28]).

As shown in Fig. 3.9, a cylinder with a diameter D_s falls with gravitational acceleration g in a stationary fluid filling a channel between parallel plates. The width and length of the channel are $H = 4D_s$ and $L = 40D_s$, respectively. The left and right walls of the channel are stationary isothermal walls maintained at constant temperature $T = T_f$, and the upper and lower walls are stationary adiabatic walls. The cylinder maintains a constant

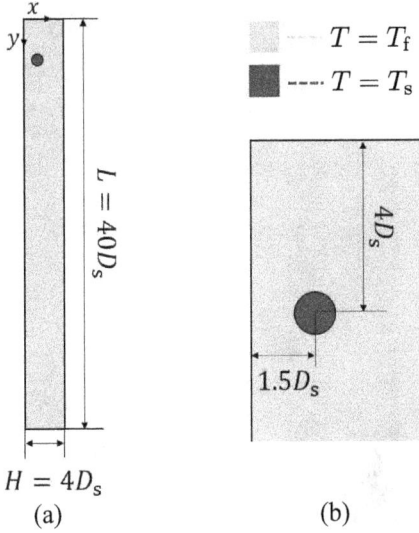

Fig. 3.9   Computational domain for sedimentation of a cold circular cylinder in a hot channel: (a) whole domain; (b) initial position of the circular cylinder.

temperature $T = T_s$ (isothermal condition), where $T_f > T_s$. The fluid is initially at rest and its initial temperature $T = T_f$ equals the wall temperature. The center of the cylinder is initially located at $(X_c, Y_c) = (1.5D_s, 4D_s)$. Note that the center of the cylinder deviates by $0.5D_s$ from the center of the channel (Fig. 3.9(b)). A cylinder of density ρ_s has a mass of $M = \rho_s(\pi D_s^2/4)$, from which the inertia moment is determined as $I = MD_s^2/8$. Also, the net gravitational force acting on the cylinder is $F_g = (1 - \rho_f/\rho_s)Mg$. In this problem, the buoyancy F_b per unit volume due to thermal expansion of the fluid is given by the following Boussinesq approximation:

$$F_b(\boldsymbol{x}, t) = \rho_f \beta g[T(\boldsymbol{x}, t) - T_f], \tag{3.53}$$

where β is the coefficient of thermal expansion.

The governing parameters of this system are the density ratio $\gamma = \rho_s/\rho_f$, the Reynolds number $\mathrm{Re} = U_{\mathrm{ref}} D_s/\nu$ (where $U_{\mathrm{ref}} = \sqrt{\pi(D_s/2)g(\gamma - 1)}$ is the reference speed), the Prandtl number $\mathrm{Pr} = \nu/\alpha$, and the Grashof number $\mathrm{Gr} = g\beta\Delta T D_s^3/\nu^2$ (where $\Delta T = T_f - T_s$ is the characteristic temperature difference). In this simulation, we set $\gamma = 1.00232$, $\mathrm{Re} = 40.5$, and $\mathrm{Pr} = 0.7$, and investigate the motion of the cylinder in three cases with different Grashof numbers $\mathrm{Gr} = 564$, 2000, and 4500. The computational conditions

(a)   Gr = 564   Gr = 2000   Gr = 4500

Temperature

(b)

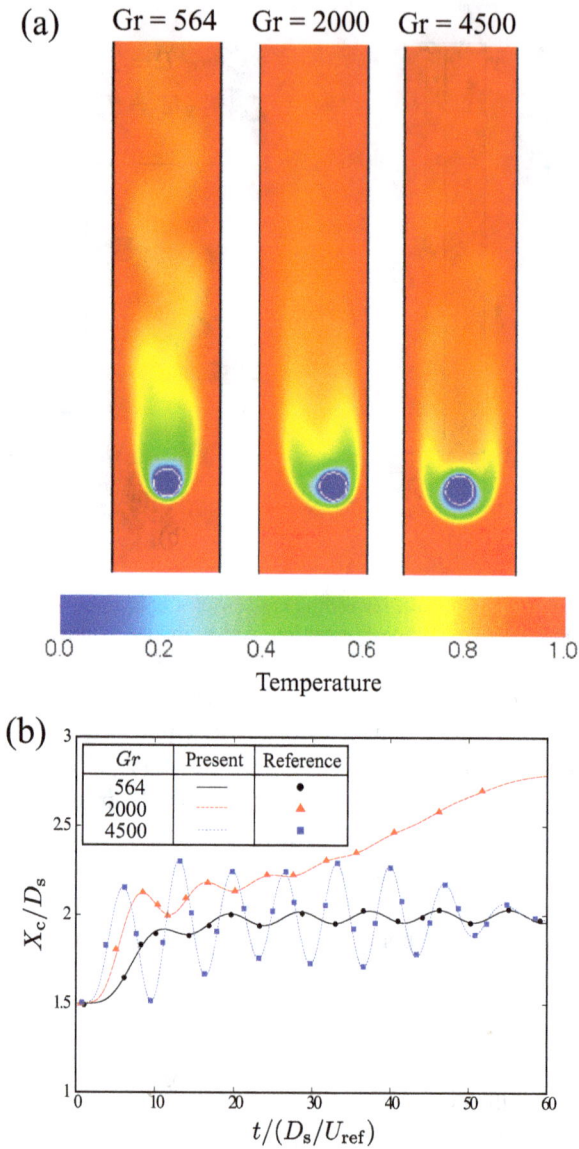

Fig. 3.10   (a) Temperature fields around a circular cylinder (in the temperature representation $(T - T_s)/(T_f - T_s)$); (b) time variations of the horizontal position X_c of the circular cylinder [138]. For comparison, the results of Eshghinejadfard and Thévenin [28] are also shown. Reprinted from Figs. 6 and 7 in [138] with permission from Elsevier.

are set as follows: $D_s = 60\Delta x$, $N_b = 244$, $U_{\text{ref}} = 0.00675$, $\tau_f = 0.5300$, and $\tau_g = 0.5428$. In addition, the internal mass effect is computed using the Lagrangian points approximation introduced in Sec. 3.5.

Figure 3.10 shows the temperature field around the cylinder at a certain moment and the time variation of the horizontal position X_c of the circular cylinder. As clarified in Fig. 3.10(a), the horizontal position of the cylinder and the temperature field in the wake greatly depend on the Grashof number. Changing the Grashof number changes the buoyancy force due to the thermal expansion of the fluid, and consequently the movement of the cylinder. Obviously, the temperature field, flow field, and object motion interact complicatedly in this problem. Also, as found in Fig. 3.10(b), the cylinder oscillates around the central axis of the channel ($x = 2D_s$) when Gr = 564 but drifts far from the central axis when Gr = 2000. When the Grashof number further increases to Gr = 4500, the cylinder returns to an oscillatory motion around the central axis of the channel. That is, these results are complicated and interesting phenomena that show a non-monotonous tendency with respect to the Grashof number. The present results well agree with those of Eshghinejadfard and Thévenin [28], confirming the validity of the thermal IB-LBM.

3.10.3   *Forward flight of a butterfly*

Insects freely fly in the air by flapping their wings. The airflows induced by the wing-flapping provide large lift and thrust forces. Therefore, the flapping flight of insects is another typical example of moving boundary flows. The authors have been performing numerical simulations of flapping flight using a simple wing model that imitates insects. Our aim is to clarify the general properties that do not depend on the specific differences among individual insects. Here we introduce the results of the study of a three-dimensional flapping wing–body model [136] that imitates a butterfly.

The butterfly-like flapping wing–body model (see Fig. 3.11) is a simple model composed of two thin wings and one thin rod-shaped body. This model generates lift and thrust forces by flapping its wings downward and backward, respectively. The wing motion is expressed as a combination of the flapping angle $\theta(t)$ and the angle of attack $\alpha(t)$, respectively defined as follows:

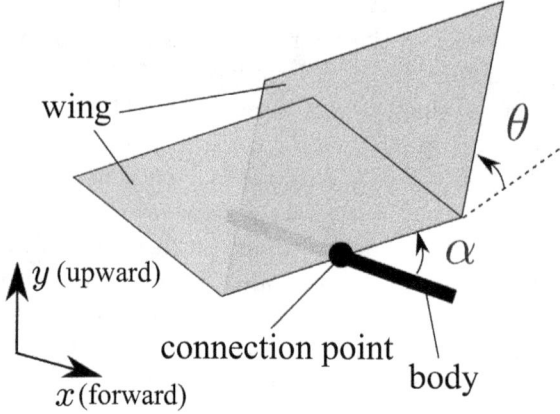

Fig. 3.11 Butterfly-like flapping wing–body model [136].

$$\theta(t) = \theta_m \cos\left(\frac{2\pi}{T}t\right), \tag{3.54}$$

$$\alpha(t) = \frac{\alpha_m}{2}\left[1 + \cos\left(\frac{2\pi}{T}t + \phi\right)\right], \tag{3.55}$$

where θ_m is the flapping amplitude, T is the period of the flapping motion, α_m is the maximum angle of attack, and ϕ is the phase difference between the flapping angle and the angle of attack. Here, we set $\theta_m = 45°$, $\alpha_m = 90°$, and $\phi = \pi/2$. Note that in this simple model, we neglect the mass and flexibility of the wing. The effects of the wing mass and flexibility on the aerodynamic performance have recently been investigated in [139, 140].

We define the characteristic length by $L_{\mathrm{ref}} = \sqrt{S}$, where S is the area of one wing, and the characteristic speed by $U_{\mathrm{ref}} = 4\theta_m L_{\mathrm{ref}}/T$. The speed U_{ref} is the mean flapping speed at the point apart from the wing bottom by L_{ref}. The governing parameter of the fluid motion is the Reynolds number defined by $\mathrm{Re} = U_{\mathrm{ref}} L_{\mathrm{ref}}/\nu$. The governing parameters of the butterfly motion are the non-dimensional mass $N_M = M/(\rho_f L_{\mathrm{ref}}^3)$ (where M is the mass of the model and ρ_f is the fluid density) and the Froude number $\mathrm{Fr} = U_{\mathrm{ref}}/\sqrt{gL_{\mathrm{ref}}}$ (where g is the gravitational acceleration). The lift coefficient C_L and thrust coefficient C_T, which give the aerodynamic performance of the model, are respectively defined as follows:

$$C_L = \frac{F_{\mathrm{lift}}}{0.5\rho_f U_{\mathrm{ref}}^2(2S)}, \qquad C_T = \frac{F_{\mathrm{thrust}}}{0.5\rho_f U_{\mathrm{ref}}^2(2S)}, \tag{3.56}$$

where F_{lift} and F_{thrust} are the lift (vertical upward) and thrust (forward) forces, respectively.

The computational domain is a rectangular domain of width W, height H, and depth H. Periodic boundary conditions are imposed on the two sides perpendicular to the x-axis, and the no-slip condition is imposed on the remaining sides. This condition represents the physical situation of the butterfly flapping in a long duct, in which W is sufficiently longer than L_{ref}. Initially, the model is placed at the center of the domain and the fluid is in a stationary state. To reduce the computational cost, this simulation employs the multi-block grid method introduced in Sec. 3.9. The fine grid is used only within a small rectangular domain containing the entire model, whose center coincides with the center of mass of the model. This grid moves along with the model. The remainder of the domain is simulated on a stationary coarse grid. The computational load is further reduced by dividing the computational domain into two equal parts in the z-direction and imposing the specular condition on the cross section. In other words, we assume symmetry of the flow field with respect to the x–y plane. In this setup, the translational motion of the body is restricted in the x- and y-directions and the rotational motion is restricted to pitching motions. Note that as this model has no volume, the internal mass effect is ignored.

We first show a case where the wing planform is square and the body of the model is fixed. In this case, the domain size is $W = H = 12L_{ref}$, the inner fine grid is sized $2.4L_{ref} \times 2.4L_{ref} \times 2.4L_{ref}$, the characteristic length is $L_{ref} = 60\Delta x$, and the Reynolds number is Re = 500. Figure 3.12 shows the time variations of the lift and thrust coefficients and the vortex structure around the model. The vortex structure is visualized by the isosurfaces of the Q criterion [55]. The time variations of the lift and thrust forces present large positive peaks during the downstroke and upstroke, respectively. These large positive lift and thrust forces drive the model upward and forward. As seen in Fig. 3.12(b), the strong vortices are attached to the leading edge and the wing tip during the downstroke and upstroke. These vortices, known as the leading-edge vortex and the wing-tip vortex, are regarded as major sources of lift and thrust generation in flapping flight. Note also that after released from the wings, the vortices are convected downward and backward. Similar vortex structures have been observed in experiments using actual butterflies [37] and in calculations using a numerical model that precisely reproduces the flight of real butterflies [160]. Although the model body is fixed in this case, forward and upward motions of the model against gravity [136] have been confirmed in free-flight simulations using actual butterfly parameters [26] (Re = 1190, N_M = 3.36, Fr = 2.35).

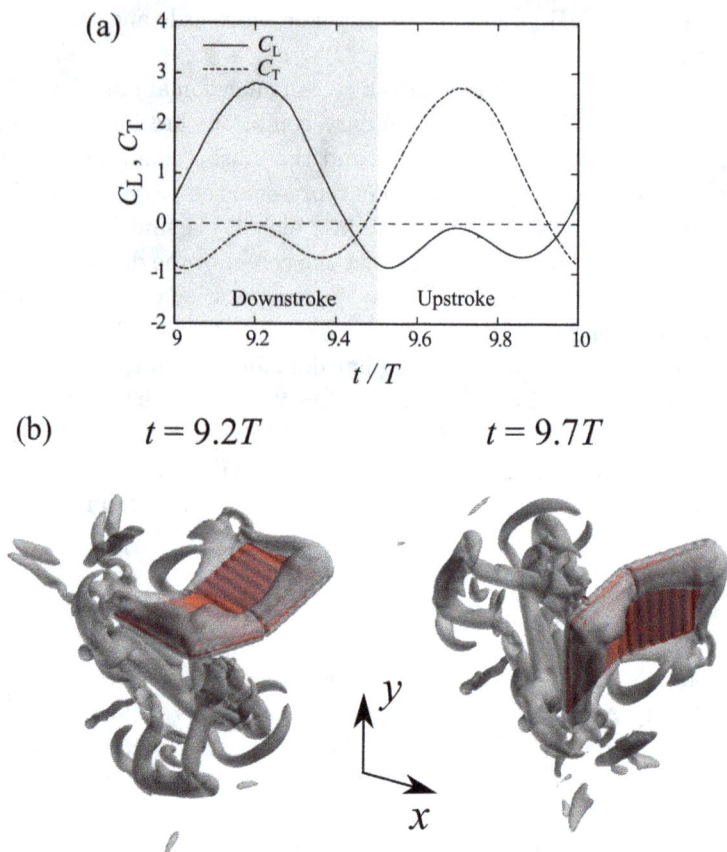

Fig. 3.12   (a) Time variations of the lift coefficient C_L and thrust coefficient C_T; (b) vortex structure around the model in a case where the wing planform is square and the body of the model is fixed [137]. In (b), the vortex structure is visualized by isosurfaces of the Q criterion ($Q = 20(U_{ref}/L_{ref})^2$). Reprinted from Fig. 6 in [137] with permission from Global Science Press.

Next, we simulate a model with the wing planform of an actual butterfly (*Janatella leucodesma*) in free flight (without pitching motion). In this case, the domain size is $W = 18L_{ref}$ and $H = 12L_{ref}$, the inner fine grid is sized $4.5L_{ref} \sin\theta_m \times 4.5L_{ref} \sin\theta_m \times 4.5L_{ref}$, and the characteristic length is $L_{ref} = 60\Delta x$. The Reynolds number is Re = 500, the non-dimensional mass is $N_M = 5.15$, and the Froude number is Fr = 2.19. The values of N_M and Fr are those of the actual butterfly, but the Reynolds number is reduced to save the computational time. Figure 3.13 shows the vortex structure around the model, the trajectory of the body center, and the time variation of the

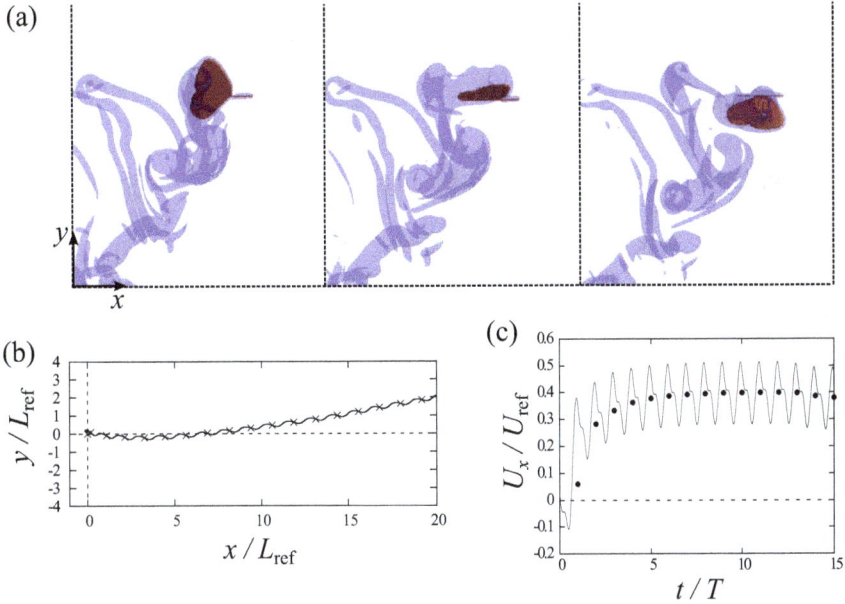

Fig. 3.13 (a) Vortex structure around the model at $t/T = 5.0$ (left), 5.2 (middle), and 5.4 right); (b) trajectory of the body center; (c) time variation of the forward speed U_x. In (a), the vortex structure is visualized by the vorticity isosurfaces ($|\nabla \times \boldsymbol{u}| = 4U_{\text{ref}}/L_{\text{ref}}$). The solid circles ($\bullet$) in (c) denote the time-averaged value of U_x in each period. Reprinted from Fig. 15 in [137] with permission from Global Science Press.

forward speed U_x. The model is observed to move upward and forward while releasing the vortices downward and backward, respectively. Note also that the forward speed varies significantly during one period, and its time-averaged value in each period converges to approximately $U_x/U_{\text{ref}} = 0.4$. This speed is much smaller than the forward speed of actual butterflies $U_x/U_{\text{ref}} = 0.88$ [26], indicating that although the model can generate enough lift force to support the actual butterfly's weight, it cannot generate enough thrust force to achieve the forward speed of $U_x/U_{\text{ref}} = 0.88$. Obtaining a sufficient thrust force requires further investigation.

3.10.4 *Sharp-turn flight of a dragonfly*

We now show a numerical example of dragonfly flight. Dragonflies differ both morphologically and functionally from other insects. By using both their forewings and hindwings, dragonflies achieve excellent flight performance with stable behaviors such as hovering and excellent maneuverability such as sharp turns and sudden starts. In a free-flight simulation of a dragonfly-like flapping wing–body model, the authors investigated how the phase difference between the forewings and hindwings influences the flight direction, lift, and thrust of the model. Within this model, both hovering and target flights can be realized by controlling the pitching rotational motion with the lead–lag angle and by changing the stroke angle of flapping and the forewing–hindwing phase difference. Furthermore, we achieved turning flight in a free-flight simulation with six degrees of freedom (three degrees of freedom each for the translational and rotational motions). Here we present an example only; the details are provided in [50, 51, 97].

Figure 3.14 shows our three-dimensional flapping wing–body dragonfly

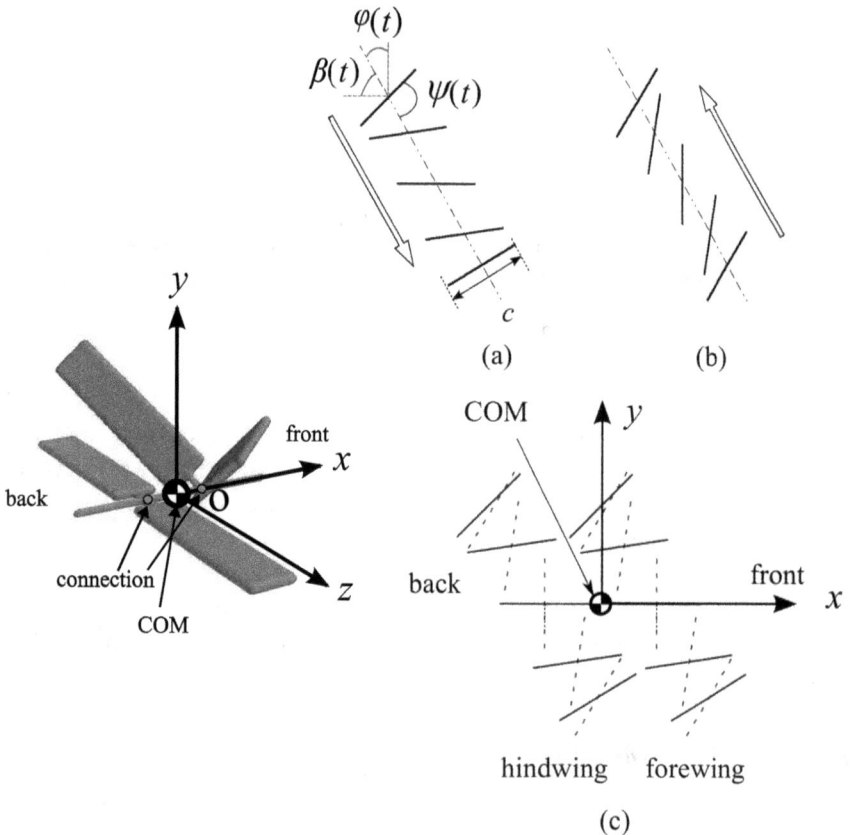

Fig. 3.14   A three-dimensional flapping wing–body model of a dragonfly (left) and its wing motions (right): (a) downstroke; (b) upstroke; (c) forewing and hindwing motions.

model. This wing–body model consists of four wings and a body. The wing mass is ignored because it is negligibly smaller than the body mass in real dragonflies. Therefore, the center of mass of the wing–body model coincides with the center of mass of the body, and its mass is denoted by M. The four wings have the same shape and do not deform. They are rectangular with a very thin short side c and long side $L = 4.5c$ (wing area $S = cL$). The body is a uniform-density rod of length $L_b = 5c$. The wings and body are connected by a very thin rod of length $0.5c$. As the center of mass of an actual dragonfly is near the midpoint of the connection between the forewings and hindwings, the forewings and hindwings of the model are connected to the body at $x = 0.75c$ and $-0.75c$, respectively. The wings in this model simultaneously perform flapping, feathering, and lead–lag motions (right figure of Fig. 3.14). Note that the wing is almost horizontal during the downstroke, and upright during the upstroke. Here, the lead–lag angle $\gamma(t)$ is the angle around the y-axis, the flapping angle $\theta(t)$ is the angle between the average chord of the wing and the z-axis, and the feathering angle (angle of attack) $\psi(t)$ is the angle between the stroke plane and the wing cross section. The stroke angle $\beta(t)$ is the angle between the x–z plane and the stroke plane. The flapping period is denoted by T. The governing parameters of this problem are the Reynolds number $\mathrm{Re} = u_{\max}c/\nu$ (where $u_{\max}$ is the maximum velocity located at $10c/3$ from the root of the wing), the Froude number $\mathrm{Fr} = u_{\max}/\sqrt{gc}$ (where g denotes gravitational acceleration), and the non-dimensional mass $m = M/(4\rho_f cS)$ (where ρ_f is the fluid density). From the data of an actual dragonfly (*Aeshna juncea*, Common hawker) [132], namely, $\hat{M} = 754$ mg, $\hat{c} = 8.1$ mm, $\hat{L} = 47$ mm, $\hat{T} = 1/36$ s, $\hat{\theta}_0 = 34.5°$, and $\hat{u}_{\max} = 4.27$ m/s, the governing parameters are estimated $\mathrm{Re} = 2300$, $\mathrm{Fr} = 15$, and $m = 51$.

We first show the calculated results in free flight with two translational degrees of freedom (forward and ascending motion) and one rotational degree of freedom (pitching motion). The calculation is performed in one-half of the region, applying the specular condition on the surface of $z = 0$. The computational domain is sized $25c \times 25c \times 10c$. To reduce the computational load, we apply the multi-block grid method. The grid spacing of the fine grid that moves with the wing–body model is Δx, and that of the stationary coarse grid is $2\Delta x$. The moving fine grid is sized $12c \times 12c \times 6c$ ($288\Delta x \times 288\Delta x \times 144\Delta x$). To reduce the effect of the initial condition with a stationary fluid, the calculation of the equation of motion of the body is started at $t = 3T$. We also investigate the Reynolds-number dependence of the aerodynamic coefficients on grids with different widths: $c = 20\Delta x$ for

Re = 40, $c = 25\Delta x$ for Re = 200, and $c = 75\Delta x$ for Re = 600. Figure 3.15(a) plots the time variations of the lift coefficient C_L, thrust coefficient C_T, and moment coefficient C_M, and Fig. 3.15(b) shows the vorticity isosurfaces at the different Reynolds numbers. The phase difference between the flapping motions of the forewings and hindwings is $\phi = 90°$ (led by the hindwings). The coefficients C_L, C_T, and C_M are respectively defined as follows:

$$C_L = \frac{F_{lift}}{\frac{1}{2}\rho_f u_{max}^2 4S}, \qquad C_T = \frac{F_{thrust}}{\frac{1}{2}\rho_f u_{max}^2 4S}, \qquad C_M = \frac{N}{\frac{1}{2}\rho_f u_{max}^2 4Sc}, \qquad (3.57)$$

where F_{lift} and F_{thrust} are the lift and thrust forces, respectively, and N is the pitching moment. Examining the time variations of the aerodynamic coefficients, we find that at all Reynolds numbers, C_L becomes positive during the downstroke ($4 \le t/T \le 4.5$) and C_T is positive during the upstroke ($4.5 \le t/T \le 5$). These forces drive the upward and forward motions of the model. Turning to the Reynolds-number dependence of the aerodynamic coefficients, we find good agreement between the results of Re = 200 and 600, and a different result for Re = 40. However, smaller vortices are generated in the wake when Re = 600 than when Re = 200. These numerical results show different flow fields in the range 100 < Re < 1000, although the aerodynamic coefficients hardly change. When Re = 200 and $\phi = 0°$, the forward velocity $\hat{U}_x$ becomes $\hat{U}_x = 0.7$ m/s, which almost corresponds to the forward speed of real dragonflies (0.7 to 3.2 m/s [6]).

Finally, we demonstrate turning flight in the free-flight simulation with three translational degrees of freedom and three rotational degrees of freedom. To conserve computational time, we set the fluid viscosity to 11.5 times that of air, giving Re = 200. The computational domain and moving fine grid are sized $25c \times 30c \times 40c$ and $12c \times 12c \times 12c$ ($288\Delta x \times 288\Delta x \times 288\Delta x$), respectively. Figure 3.16 shows the numerical results of the turning flight. By controlling the stroke angle and flapping period of the left and right wings, the dragonfly turns through approximately 180°. The turning radius is almost twice the wing length and the turning time is short (around 0.36 seconds). Such sharp turns are characteristic features of real dragonfly flight.

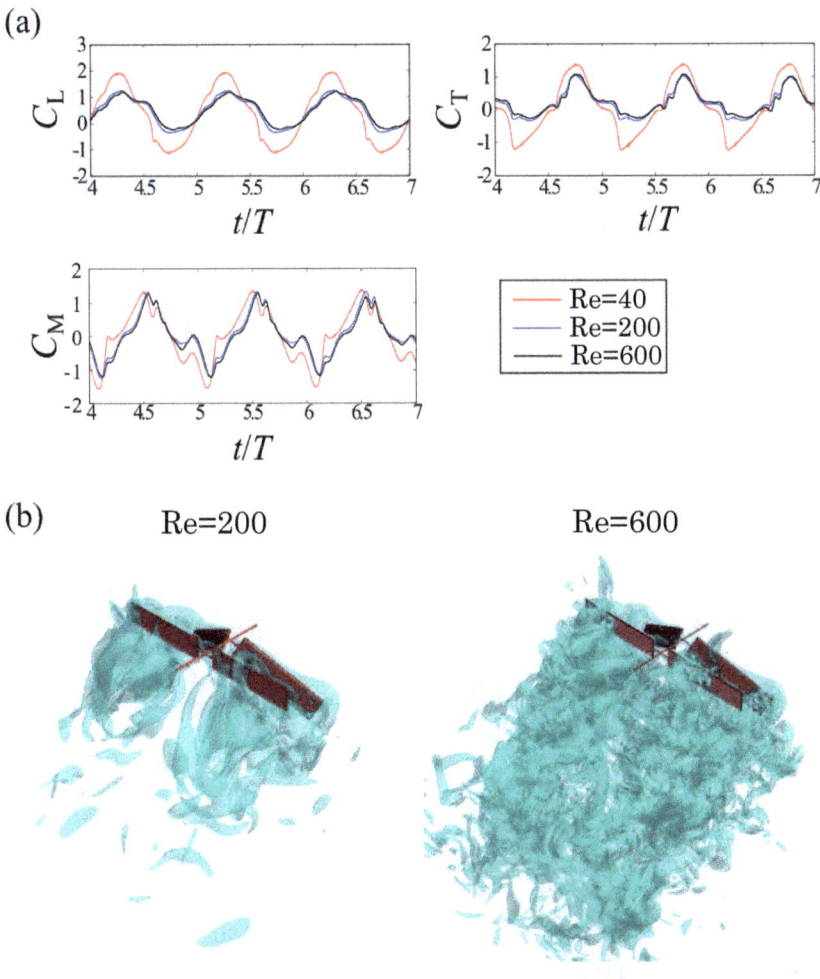

Fig. 3.15   (a) Time variations of aerodynamic coefficients C_L, C_T, and C_M in the dragonfly model; (b) vorticity isosurfaces ($|\nabla \times \boldsymbol{u}|c/u_{\max} = 1.0$, $t = 5.75T$). Reprinted from Fig. B1 in [97] with permission from IOP Publishing.

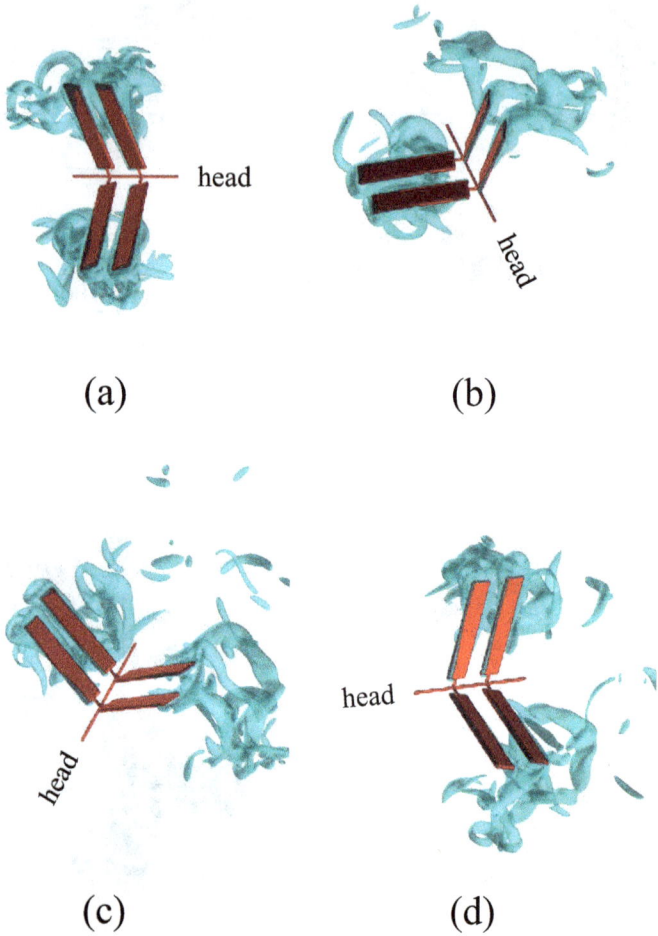

Fig. 3.16 Isosurfaces of the vorticity around flapping wings during a turn of approximately 180° ($|\nabla \times \boldsymbol{u}|c/u_{\mathrm{max}} = 1.5$): (a) $t = 3T$; (b) $t = 8T$; (c) $t = 12T$; (d) $t = 16T$. Reprinted from Fig. 10 in [50] with permission from IOP Publishing.

Chapter 4

Two-Phase Lattice Boltzmann Method (Two-Phase LBM)

This chapter introduces the two-phase LBMs, which are applicable to liquid–liquid and gas–liquid two-phase flows. Two-phase flows are characterized by the existence of freely deformable interfaces. If the interface is regarded as a two-phase boundary, a two-phase flow can be considered as a moving boundary flow. Numerical calculations of two-phase flows are difficult because they must capture sharp interfaces, track the interfaces at every moment, and guarantee the conservation of mass of each phase. As described in Chap. 1, the LBM is a mass-conserving method. Studies on two-phase LBMs began in the mid-1990s, shortly after the LBM was proposed. Initially, the density ratio in two-phase LBMs was limited to around 10, but two-phase LBMs that handle density ratios of approximately 1000 have been proposed since 2004–2006 [69, 91, 166], and two-phase LBMs have been applied in a wide range of fields. Many two-phase LBMs with a high density ratio have also been proposed. Instead of introducing these methods comprehensively, this chapter explains the essential issues of two-phase LBMs with high density ratios and introduces recent methods proposed by the authors.

4.1 Classification of Two-Phase LBM

Two-phase LBMs can be roughly classified into three methods with different treatments of the two phases (gas and liquid phases).

(1) Color-gradient model

This model evolved from the lattice gas model (an original particle model that tracks collisions and translations by placing fictitious particles on lattice points instead of sequentially calculating the velocity distribution function of fictitious particles) and was introduced into the LBM by

Gunstensen et al. [41]. This model considers two color-coded particles that interact in an appropriate way to represent the interface. The color-gradient model is conceptually similar to the VOF method, which is often used in numerical calculations of two-phase flows. Initially, the method was limited to two-phase flows with equal density, but has since evolved to higher density ratios [54]. In one numerical example involving stationary droplets and bubbles, the density ratio reached 10000, but in the calculation of moving droplets and bubbles, the highest stable density ratio is approximately 10.

(2) Pseudo-potential model

Since its proposal by Shan and Chen [122], the pseudo-potential model has been widely published. This model replaces the color-gradient model with a phenomenological physical model that expresses the interface as a pseudo-potential between the particles. The pseudo-potential can be obtained from the equation of state representing the two phases (e.g., the van der Waals equation). This pseudo-potential is added as an external force term in the time evolution equation of the velocity distribution function. In the initial formulation, the density ratio was limited to around 10, but later researchers devised a function form of the equation of state or increased the number of particle velocities [16, 94]. These formulations increased the possible density ratio to around 1000. However, a simulation study of two-phase Poiseuille flows reported that velocity discontinuity occurs at density ratios above 20 [100]. More recently, a method based on the non-orthogonal MRT-LBM has allowed calculations at a density ratio of 1000 [31].

(3) Free-energy model (Phase-field model)

First introduced to two-phase LBMs by Swift et al. [143, 144], this model forms two-phase interfaces based on non-equilibrium thermodynamics. The free-energy model is introduced as a pressure tensor in the local equilibrium distribution function. It can also be introduced as an external force term in the time evolution equation of the velocity distribution function. The same model, also called the phase-field model, is used in the calculation of interface phenomena in materials science. Like the pseudo-potential model, the free-energy model determines the densities ρ_G and ρ_L of the gas and liquid phases, respectively, from the equation of state (e.g., the van der Waals equation). The free-energy model is superior to the other two models because its interface representation is based on the non-equilibrium thermodynamics, but in calculations of high density ratios, the pressure computation tends to diverge. Therefore, a stable computation of the pressure must be developed for this method. The method of Mazloomi et al. [95] handles density ratios of approximately 100, but the pressure computation reportedly becomes unstable at density ratios above 100. More recently, a method that handles two-phase flows at a density ratio of 1000 has been proposed [29, 104, 123]. In [29], the numerical stability is enhanced by improving the MRT model and devising a finite-difference scheme. Sisompul and Aoki [123] stabilized a cumulant collision model by filtering (spatial averaging) the pressure and velocity fields. Otomo et al. [104] developed the method in [29] to simplify the computational scheme using the regularized LBM.

Above we introduced three interface models, each with its advantages and disadvantages. All models avoid the need for explicitly tracking the time variations of the interface shape, and all have excellent mass conservation. The following describes the details of the free-energy model used in this book.

4.2 Free-Energy Model

Let the order parameter ϕ be divided into ϕ_1 and ϕ_2.[1] The free energy Ψ (called the Ginzburg–Landau free energy) of the entire system, in which the two phases ϕ_1 and ϕ_2 coexist, can be expressed as follows [101, 115]:

$$\Psi[\phi] = \int \left[\psi(T, \phi) + \frac{\kappa}{2} |\nabla \phi|^2 \right] d\boldsymbol{x}, \qquad (4.1)$$

where the integration is performed over the entire region. The first and second terms of the integrand indicate the free energies of the bulk and interface, respectively. The parameter κ determines the thickness and tension at the interface.[2] The free energy ψ in the bulk (for example, in a van der Waals fluid) is given by

$$\psi(T, \phi) = \phi T \ln \left(\frac{\phi}{1 - b\phi} \right) - a\phi^2. \qquad (4.2)$$

In the above expression, T, a, and b are arbitrary constants that determine the function form of ϕ. Note that Eq. (4.2) is a double-well function.

---

[1] For example, consider that ϕ is the density ρ.

[2] In the calculations after Sec. 4.3, we define κ_f for determining the interface thickness and κ_g for determining the interfacial tension.

Now, according to the thermodynamics of a non-equilibrium system, the chemical potential μ_c is determined as

$$\mu_c = \frac{\delta \Psi}{\delta \phi} = \frac{\partial \psi}{\partial \phi} - \kappa \frac{\partial^2 \phi}{\partial x_\alpha^2}. \tag{4.3}$$

When the two values of ϕ are in equilibrium, they must have the same chemical potential in the bulk. Therefore, two stable values of ϕ (having equal $\partial \psi / \partial \phi$) are obtained as the points of contact of a common tangent of Eq. (4.2). For example, for $a = 1$, $b = 1$, and $T = 2.93 \times 10^{-1}$, we get $\phi_1 = 2.638 \times 10^{-1}$ and $\phi_2 = 4.031 \times 10^{-1}$; for $a = 1$, $b = 6.7$, and $T = 3.5 \times 10^{-2}$, we get $\phi_1 = 1.134 \times 10^{-2}$ and $\phi_2 = 9.714 \times 10^{-2}$.

A smooth interface is determined by autonomous deformation such that the free energy of the system (Eq. (4.1)) is minimized. The first term of Eq. (4.1) is minimized by a stepwise separation into the two phases ϕ_1 and ϕ_2 at the interface, whereas the second term is minimized when the interface thickness is infinitely large. Balancing both terms, we obtain a finitely thick interface with a smooth change of ϕ across the interface. The pressure tensor in the two-phase system is then obtained as follows [158]:

$$P_{\alpha\beta} = p\, \delta_{\alpha\beta} + \kappa \frac{\partial \phi}{\partial x_\alpha} \frac{\partial \phi}{\partial x_\beta}, \tag{4.4}$$

where

$$p = p_0 - \kappa \phi \nabla^2 \phi - \frac{\kappa}{2} |\nabla \phi|^2, \tag{4.5}$$

and

$$p_0 = \phi \frac{\partial \psi}{\partial \phi} - \psi = \phi T \frac{1}{1 - b\phi} - a\phi^2. \tag{4.6}$$

The above equation is nothing less than the equation of state relating the pressure p_0 to the density ρ of a van der Waals fluid (regarding ϕ as the density ρ). In addition, the flux J_α of the order parameter ϕ is determined by

$$J_\alpha = -M_\phi \phi \frac{\partial \mu_c}{\partial x_\alpha} = -M_\phi \frac{\partial P_{\alpha\beta}}{\partial x_\beta}, \tag{4.7}$$

where M_ϕ is a constant called the mobility.

To evolve ϕ in time based on the free energy, many studies adopt the Cahn–Hilliard equation [12, 13] or the conservative Allen–Cahn equation [17, 39]. These equations are discussed below.

(1) Cahn–Hilliard equation

This is a conservative model, and the following advection–diffusion equation is derived from the conservation law of ϕ:

$$\mathrm{Sh}\frac{\partial \phi}{\partial t} + \frac{\partial}{\partial x_\alpha}(\phi u_\alpha) = -\frac{\partial J_\alpha}{\partial x_\alpha} = \frac{\partial}{\partial x_\alpha}\left(M_\phi \frac{\partial P_{\alpha\beta}}{\partial x_\beta}\right). \tag{4.8}$$

The Cahn–Hilliard equation excellently conserves ϕ throughout the entire system, but the speed of movement between the two phases is large. The moving speed depends on the magnitude of the mobility M_ϕ and the function form of the free energy Eq. (4.2). Thus, small droplets and bubbles with diameters of less than $10\Delta x$ quickly disappear and dissolve into the other surrounding phase [165]. To capture these small droplets and bubbles, simulations must be performed on a very fine grid.

(2) Conservative Allen–Cahn Equation

The original Allen–Cahn equation [3] is a non-conservative model. The following equation assumes that ϕ changes proportionally to μ_c:

$$\mathrm{Sh}\frac{\partial \phi}{\partial t} + \frac{\partial}{\partial x_\alpha}(\phi u_\alpha) = -M'\mu_c, \tag{4.9}$$

where M' is a proportionality constant. Note that this equation does not conserve ϕ. The conservation of ϕ is satisfied by the following conservative Allen–Cahn equation [17, 39]:

$$\mathrm{Sh}\frac{\partial \phi}{\partial t} + \frac{\partial}{\partial x_\alpha}(\phi u_\alpha) = \frac{\partial}{\partial x_\alpha}\left[M_\phi\left(1 - \frac{1}{|\nabla\phi|}\frac{1-4\phi^2}{W}\right)\frac{\partial \phi}{\partial x_\alpha}\right], \tag{4.10}$$

where $M_\phi = M'\kappa$ and $W = \sqrt{16\kappa}$. Here we use the following simple double-well form of the free-energy function $\psi(\phi)$:

$$\psi(\phi) = \frac{1}{2}\phi^2\left(\phi^2 - \frac{1}{2}\right). \tag{4.11}$$

The chemical potential μ_c is then given by

$$\mu_c = 2\phi\left(\phi + \frac{1}{2}\right)\left(\phi - \frac{1}{2}\right) - \kappa\frac{\partial^2 \phi}{\partial x_\alpha^2}. \tag{4.12}$$

As clarified in the above equation, the two stable values of ϕ, at which both bulk chemical potentials equal μ_c, are $\phi_1 = -1/2$ and $\phi_2 = 1/2$.

The conservative Allen–Cahn equation retains small droplets and bubbles with diameters of less than $5\Delta x$ for a sufficiently long time, but numerical errors in the bulk induce small fluctuations of ϕ, which give rise to tiny droplets and bubbles that make non-physical movements.

4.3 Computation of ϕ by LKS

The above ϕ can be easily evolved in time using the LKS described in Sec. 2.1. In the following formulations, we employ the three-dimensional fifteen-velocity model.

(1) Cahn–Hilliard equation
The time evolution of ϕ is computed by

$$\phi(\boldsymbol{x}, t + \Delta t) = \sum_{i=1}^{15} f_i^{\mathrm{eq}}(\boldsymbol{x} - \boldsymbol{c}_i \Delta x, t), \tag{4.13}$$

where the local equilibrium distribution function f_i^{eq} is given by [143, 144]

$$f_i^{\mathrm{eq}} = H_i \phi + F_i \left(p_0 - \kappa_f \phi \nabla^2 \phi - \frac{\kappa_f}{6} |\nabla \phi|^2 \right) + 3 E_i \phi c_{i\alpha} u_\alpha$$
$$+ E_i \kappa_f G_{\alpha\beta}^\phi c_{i\alpha} c_{i\beta} + E_i C \left(\frac{\partial P_{\alpha\beta}}{\partial x_\beta} \right) c_{i\alpha} \Delta x, \tag{4.14}$$

and

$$\begin{cases} H_1 = 1, \quad H_2 = H_3 = \cdots = H_{15} = 0, \\ F_1 = -7/3, \quad F_i = 3 E_i, \quad i = 2, 3, \cdots, 15, \end{cases} \tag{4.15}$$

$$p_0 = \phi T \frac{1}{1 - b\phi} - a\phi^2, \tag{4.16}$$

$$G_{\alpha\beta}^\phi = \frac{9}{2} \frac{\partial \phi}{\partial x_\alpha} \frac{\partial \phi}{\partial x_\beta} - \frac{3}{2} \frac{\partial \phi}{\partial x_\gamma} \frac{\partial \phi}{\partial x_\gamma} \delta_{\alpha\beta}. \tag{4.17}$$

Note that the parameter $\kappa_f = O((\Delta x)^2)$ determines the thickness of the interface and $C = O(1)$ is a constant parameter that adjusts the mobility M_ϕ, given by

$$M_\phi = \left(\frac{1}{2} - \frac{1}{3} C \right) \Delta x. \tag{4.18}$$

In addition, $\nabla\phi$ and $\nabla^2\phi$ in Eqs. (4.14) and (4.17) are calculated by the following difference approximations:

$$\nabla\phi \approx \frac{1}{10\Delta x} \sum_{i=1}^{15} c_i \phi(x + c_i \Delta x), \tag{4.19}$$

$$\nabla^2\phi \approx \frac{1}{5(\Delta x)^2} \left[\sum_{i=2}^{15} \phi(x + c_i \Delta x) - 14\phi(x) \right]. \tag{4.20}$$

(2) Conservative Allen–Cahn equation

In the conservative Allen–Cahn equation, the time evolution of ϕ is computed by

$$\phi(x, t + \Delta t) = \sum_{i=1}^{15} \left\{ f_i^{\text{eq}}(x - c_i \Delta x, t) + AE_i[\phi(x, t) - \phi(x - c_i \Delta x, t)] \right\}, \tag{4.21}$$

where

$$f_i^{\text{eq}} = E_i \phi \left(1 + 3c_{i\alpha}u_\alpha\right) + 3E_i c_{i\alpha} n_\alpha M_\phi \frac{1 - 4\phi^2}{W}. \tag{4.22}$$

In addition, we have

$$n = \frac{\nabla\phi}{|\nabla\phi|}, \tag{4.23}$$

$$M_\phi = \frac{1}{6}(1 - A)\Delta x. \tag{4.24}$$

To avoid division by zero in the calculation of n, we insert a small positive constant ε_0 (e.g., $\varepsilon_0 = 10^{-12}$) in the denominator of Eq. (4.23) as follows:

$$n \approx \frac{\nabla\phi}{|\nabla\phi| + \varepsilon_0}. \tag{4.25}$$

Figure 4.1 shows numerical results of (1) and (2). The computational domain is divided into a $96\Delta x \times 96\Delta x \times 96\Delta x$ cubic lattice. A droplet ($\phi = \phi_2$) with a diameter of $D = 40\Delta x$ is placed at the center of the domain. The computation terminates at $20000\Delta t$. In the Cahn–Hilliard equation, we set $a = 1$, $b = 1$, $T = 2.93 \times 10^{-1}$ (where $\phi_1 = 2.638 \times 10^{-1}$ and $\phi_2 = 4.031 \times 10^{-1}$), $\kappa_f = 0.06(\Delta x)^2$, and $M_\phi = 0.5\Delta x$. In the conservative Allen–Cahn equation, we set $\phi_1 = -0.5$, $\phi_2 = 0.5$ $M_\phi = (1/60)\Delta x$ ($A = 0.9$), and $W = 4\Delta x$. Clearly, both numerical results maintain a smooth interface.

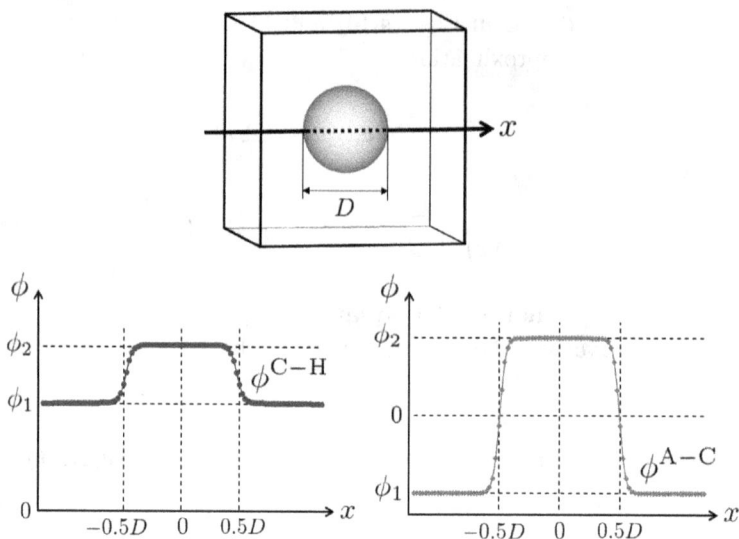

Fig. 4.1 Calculated results of ϕ by the Cahn–Hilliard equation (left) and the conservative Allen–Cahn equation (right).

4.4 Wettability on the Body Surface (Boundary Condition for ϕ)

When a body is contacted by a two-phase interface, its wettability can be changed by considering additional free energy due to the contact [10, 84]. For example, suppose that the two-phase interface at $z \geq 0$ touches the surface of the body. The gradient of ϕ in the z-direction is related to the contact angle θ_w (defined as the angle between the body surface and the two-phase interface in the liquid phase ($\phi = \phi_2$)) as follows:

$$\left(\frac{\partial \phi}{\partial z} \right)_{z=0} = \begin{cases} \text{negative}, & \theta_w < 90°, \\ 0, & \theta_w = 90°, \\ \text{positive}, & \theta_w > 90°. \end{cases} \tag{4.26}$$

As the actual relationship between $(\partial \phi / \partial z)_{z=0}$ and θ_w depends on κ_f and W, it must be found in preliminary calculations (see, e.g. [145]).

Alternatively, the surface value of ϕ can be simply given by

$$\phi_{z=0} = (1 - S)\phi_1 + S\phi_2 \qquad (0 \leq S \leq 1), \tag{4.27}$$

where the parameter S determines the wettability of the surface. Changing the value of S alters the contact angle θ_w. For example, when $S = 0$, 0.5, and 1, $\theta_w = 180$, 90, and $0°$, respectively.

4.5 Two-Phase LBM for Equal Density

First, suppose that two immiscible liquids of equal density are mixed [64]. In this case, the two fluids can be distinguished by the two values of ϕ described in Sec. 4.3, and the flow velocity and pressure can be computed using the LBM [47]. In the LBM formulation, the time evolution of the new velocity distribution function g_i is given by

$$g_i(\boldsymbol{x} + \boldsymbol{c}_i \Delta x, t + \Delta t) = g_i(\boldsymbol{x}, t) - \frac{1}{\tau_g}[g_i(\boldsymbol{x}, t) - g_i^{eq}(\boldsymbol{x}, t)]. \qquad (4.28)$$

The pressure p and flow velocity $\boldsymbol{u}$ are respectively calculated by[3]

$$p = \frac{1}{3} \sum_{i=1}^{15} g_i, \qquad (4.29)$$

$$\boldsymbol{u} = \sum_{i=1}^{15} \boldsymbol{c}_i g_i. \qquad (4.30)$$

Also, the local equilibrium distribution function g_i^{eq} is given by

$$g_i^{eq} = E_i \left(3p + 3u_\alpha c_{i\alpha} + \frac{9}{2} u_\alpha u_\beta c_{i\alpha} c_{i\beta} - \frac{3}{2} u_\alpha u_\alpha \right) + E_i \kappa_g G_{\alpha\beta}^\phi c_{i\alpha} c_{i\beta}. \qquad (4.31)$$

The last term on the right-hand side of this expression is the interfacial tension term, and $\kappa_g = O((\Delta x)^2)$ is a parameter that determines the interfacial tension.

The kinematic viscosity coefficient of the fluid ν is given by

$$\nu = \frac{1}{3} \left(\tau_g - \frac{1}{2} \right) \Delta x. \qquad (4.32)$$

Therefore, the kinematic viscosities of the two fluids can be changed by changing τ_g. Also, the interfacial tension $\sigma = O((\Delta x)^2)$ is given by [63,115]

$$\sigma = \kappa_g \int_{-\infty}^{\infty} \left(\frac{\partial \phi}{\partial \xi} \right)^2 d\xi, \qquad (4.33)$$

where ξ is the coordinate perpendicular to the interface. By changing κ_g, we can change the magnitude of interfacial tension σ over a wide range.

---

[3]At the interface, we must add $\frac{2}{3}\kappa_g |\nabla \phi|^2$ to the pressure p obtained by Eq. (4.29).

4.6 Pressure Computation at Large Density Ratio

As described in Sec. 4.1, the pressure computation in the two-phase LBM usually becomes unstable at large density ratios. The following methods can stabilize the pressure computation in the two-phase LBM at large density ratios.

(1) Method based on the continuity equation and the equation of state (thermodynamic pressure)

The density ρ (replacing ϕ) is computed in the Cahn–Hilliard equation (4.8) and substituted into the van der Waals equation of state (4.6) for computing the pressure p_0. This method is equivalent to the LBM for single-phase flows in which the equation of state is $p = c_s^2 \rho$ with $c_s = \sqrt{1/3}$. The obtained pressure is called the "thermodynamic pressure" because of using the equation of state.

In a well-devised formulation, the density change at the interface and the flow velocity in the domain can be computed using a single velocity distribution function, as in the LBM for single-phase flows. However, the speed of sound $c_s = \sqrt{dp/d\rho}$ differs between the gas and liquid phases. In some function forms of the equation of state, the speed of sound through gas becomes $c_s \ll 1$. In such cases, the limiting condition of the LBM, namely, $|u|/c_s \ll 1$ in low Mach number flows cannot be satisfied.

(2) Method based on the pressure evolution equation (hydrodynamic pressure)

Deforming the continuity equation of the compressible fluid (Eq. (4.8) or Eq. (4.10), where ρ replaces ϕ and the right-hand side is set to 0) with the equation of state $p = f(\rho)$, we get

$$\text{Sh}\frac{\partial p}{\partial t} + \rho c_s^2 \frac{\partial u_\alpha}{\partial x_\alpha} + u_\alpha \frac{\partial p}{\partial x_\alpha} = 0. \tag{4.34}$$

This equation is called the pressure evolution equation. The third term in this equation is often ignored because it is smaller than the other two terms. Considering a new velocity distribution function different from the velocity distribution function that is used for the computation of the density in the interface, and formulating the time evolution equation of the new velocity distribution function so as to satisfy Eq. (4.34), we can directly compute the pressure p from the moment of the function [91]. The speed of sound c_s in Eq. (4.34) is often assumed as $c_s = \sqrt{1/3}$, the speed of sound in single-phase flows. However, c_s can be freely chosen. The pressure p obtained from Eq. (4.34) is called the "hydrodynamic pressure." This pressure is determined independently of the two-phase equation of state and is not thermodynamically related to the density. This distinction is clarified by the use of two velocity distribution functions, one for computing the density in the domain and the other for computing the flow velocity and pressure in the domain. This is a weakness of using the pressure evolution equation. However, even in conventional methods such as the VOF, front-tracking, and level set methods, the interface representation is separated from the pressure computation. Therefore, the above treatment of two-phase flows is also followed in conventional methods.

Although the above two methods are equivalent in principle, Eq. (4.34) is known to yield more stable results [81]. In two-phase LBMs, the thermodynamic pressure computed by (1) is limited to density ratios below 100, but the hydrodynamic pressure computed by (2) can be stably computed at density ratios of approximately 1000. Also, if the density of the gas phase becomes much smaller than the density of the liquid phase, the velocity field of the gas phase will fluctuate largely. Consequently, as the condition $|\boldsymbol{u}|/c_s \ll 1$ is not necessarily satisfied and $\nabla \cdot \boldsymbol{u} \neq 0$ in the gas phase, the computation often diverges. Computing bubble flows with density ratios of approximately 1000 is particularly difficult [93]. At density ratios around 1000, we must therefore ensure that the thermodynamic or hydrodynamic pressure p satisfies the continuity equation $\nabla \cdot \boldsymbol{u} = 0$ of the incompressible fluid. For this purpose, an appropriate method is required.

Around 2005, researchers began developing two-phase LBMs that compute flows with density ratios near 1000 [69,91,166]. These two-phase LBMs have been applied in a wide range of fields. In the two-phase LBM proposed by Inamuro et al. [69], the hydrodynamic pressure must be computed using the Poisson equation. In Lee and Lin's two-phase LBM [91], the hydrodynamic pressure is computed by (2), and the numerical stability is improved by a finite-difference scheme. Meanwhile, the two-phase LBM proposed by Zheng et al. [166] assumes equal density of two phases. Various two-phase LBMs for large density ratios were introduced in Sec. 4.1, but no definitive method has yet been established, and further improvements of the existing methods are ongoing.

In the next section, we will describe a computational scheme that computes the flow velocity and hydrodynamic pressure of two-phase flows with large density ratios. The scheme adopts the improved LKS described in Sec. 2.6.

4.7 Two-Phase LBM for Large Density Ratios

This section introduces a two-phase LBM based on the improved LKS described in Sec. 2.3 [74, 76]. As the improved LKS has excellent numerical stability (see Sec. 2.7), it is considered suitable for calculating two-phase flows with large density ratios, which are difficult to compute by other schemes. In the following formulations, the densities of the gas and liquid phases are $\rho_G = 1$ and $\rho_L > 1$, respectively.

(1) Formulation

First, the two-phase density ρ is determined from the ϕ computed by the method in Sec. 4.3. For example, we can linearly interpolate as follows:

$$\rho = \frac{\phi - \phi_{\min}}{\phi_{\max} - \phi_{\min}}(\rho_L - \rho_G) + \rho_G, \tag{4.35}$$

where $\phi_{\max}$ and $\phi_{\min}$ are the maximum and minimum values of ϕ, respectively, at the given moment. Alternatively, we can express the function form of ρ at the interface:

$$\rho = \begin{cases} \rho_G, & \phi = \phi_{\min}, \\ \frac{\Delta\rho}{2}\left[\sin\left(\frac{\phi - \overline{\phi}}{\Delta\phi}\pi\right) + 1\right] + \rho_G, & \phi_{\min} < \phi < \phi_{\max}, \\ \rho_L, & \phi = \phi_{\max}, \end{cases} \tag{4.36}$$

where $\Delta\rho = \rho_L - \rho_G$, $\Delta\phi = \phi_{\max} - \phi_{\min}$, and $\overline{\phi} = (\phi_{\max} + \phi_{\min})/2$.

The viscosity coefficient μ at the interface can be obtained either by the following linear interpolation:

$$\mu = \mu_G + \frac{\rho - \rho_G}{\rho_L - \rho_G}(\mu_L - \mu_G), \tag{4.37}$$

or by the following harmonic interpolation:

$$\frac{1}{\mu} = \frac{1}{\mu_G} + \frac{\rho - \rho_G}{\rho_L - \rho_G}\left(\frac{1}{\mu_L} - \frac{1}{\mu_G}\right). \tag{4.38}$$

In these expressions, μ_G and μ_L are the viscosity coefficients of the gas and liquid phases, respectively. Both coefficients are of magnitude $O(\Delta x)$.

To compute the flow velocity and pressure in the improved LKS, we first evolve the pressure $p(\boldsymbol{x}, t)$ in time by the following procedure.

Step 0. Give the initial values of $p_0(x) = p(x,t)$.[4]
Step 1. Compute the pressure in the $(l + 1)$-th iteration:

$$p_{l+1}(x) = p_l(x) + \frac{\omega(x,t)}{3} \sum_{i=1}^{15} \left[\Delta P_{i,l} + g_i^{\text{eq}}(x - c_i \Delta x, t) \right], \quad (4.39)$$

where

$$\Delta P_{i,l} = \frac{3}{2} E_i \left[\frac{1}{\rho(x - c_i \Delta x, t)} + \frac{1}{\rho(x,t)} \right] [p_l(x - c_i \Delta x) - p_l(x)].$$
$$(4.40)$$

Perform n iterations of this process.
Step 2. Update the pressure $p(x, t + \Delta t) = p_n(x)$ at the new time.

The parameter $\omega(x,t)$ in Eq. (4.39) is determined as

$$\omega(x,t) = 1 + \frac{\rho(x,t) - \rho_{\text{G}}}{\rho_{\text{L}} - \rho_{\text{G}}}(\omega_{\text{max}} - 1), \quad (4.41)$$

where ω_{max} is a constant. We also define g_i^{eq} as the local equilibrium distribution function without a pressure term:

$$g_i^{\text{eq}} = E_i \left(3c_{i\alpha} u_\alpha + \frac{9}{2} c_{i\alpha} c_{i\beta} u_\alpha u_\beta - \frac{3}{2} u_\alpha u_\alpha \right). \quad (4.42)$$

In addition, $\Delta P_{i,l}$ is the pressure difference developed between two phases with different densities.

The parameter $\omega(x,t)$ given by Eq. (4.41) is a kind of acceleration parameter [124]. By setting the maximum value ω_{max} of this acceleration parameter and the number of iterations n to $n\omega_{\text{max}} \approx \rho_{\text{L}}$, the speed of sound becomes $c_{\text{s}} \approx \sqrt{1/3}$ (see (**2**) Asymptotic analysis (S-expansion) below). However, beyond some upper limit ω_{max}, the computations become unstable. To satisfy the condition $n\omega_{\text{max}} \approx \rho_{\text{L}}$, we must increase the number of iterations n. That is, we must iterate **Step 1** to improve the numerical stability.

Next, the time evolution of the flow velocity $u(x,t)$ is computed by

---

[4]Not to be confused with p_0 of the van der Waals equation of state (4.6).

$$u(x, t + \Delta t) = \sum_{i=1}^{15} c_i \left\{ \Delta P_i(x, t + \Delta t) + g_i^{\text{eq}}(x - c_i \Delta x, t) \right.$$

$$+ 3 A_u E_i c_i \cdot [u(x, t) - u(x - c_i \Delta x, t)]$$

$$\left. + 3 E_i c_{i\alpha} V_\alpha(x, t) \right\}$$

$$- \frac{1}{6}(1 - A_u) \lambda (\Delta x)^4 \nabla^2 (\nabla^2 u(x, t))$$

$$+ \frac{1}{\rho(x, t)} F_{\text{sv}}(x, t) \Delta x, \qquad (4.43)$$

where

$$\Delta P_i(x, t) = \frac{3}{2} E_i \left[\frac{1}{\rho(x - c_i \Delta x, t)} + \frac{1}{\rho(x, t)} \right] [p(x - c_i \Delta x, t) - p(x, t)], \qquad (4.44)$$

$$V_\alpha = \frac{1}{\rho} \frac{\partial \mu}{\partial x_\beta} \left(\frac{\partial u_\beta}{\partial x_\alpha} + \frac{\partial u_\alpha}{\partial x_\beta} \right) \Delta x. \qquad (4.45)$$

The term V_α accounts for the change of viscosity coefficient in the two phases. Recalling the improved LKS (Sec. 2.3), we find the following form of the viscosity coefficient μ:

$$\mu = \frac{1}{6} \rho (1 - A_u) \Delta x, \qquad (4.46)$$

where $A_u = O(1)$ is a constant. The term with the constant $\lambda = O(1)$ is the viscous dissipation term due to hyper-viscosity, which is added to improve the numerical stability. This viscous dissipation term dissipates the oscillations in the solution over a small scale. Its magnitude is $O((\Delta x)^4)$ and proportional to μ, so it preserves the numerical accuracy even in high Reynolds number flows. The viscosity coefficient of Eq. (4.46) have a range where we can perform stable computations. For example, in the simulation of a stationary droplet with the density ratio of 800, we have $0.001 \leq \nu/\Delta x \leq 1/6$ for $\lambda = 0$, $1 \times 10^{-4} \leq \nu/\Delta x \leq 2 \times 10^{-2}$ for $\lambda = 5$, and $5 \times 10^{-5} \leq \nu/\Delta x \leq 1 \times 10^{-2}$ for $\lambda = 10$. Finally, we obtain the body force (external force per unit volume) $F_{\text{sv}}(x, t)$ ($|F_{\text{sv}}| = O((\Delta x)^2)$) by the so-called continuous surface force (CSF) model [9], which expresses the interfacial tension σ ($= O((\Delta x)^2)$). The CSF formulation is

$$F_{\text{sv}}(x, t) = \sigma \chi(x, t) \frac{\nabla \rho(x, t)}{[\rho]} \frac{\rho(x, t)}{\langle \rho \rangle}, \qquad (4.47)$$

where

$$[\rho] = \rho_{\rm L} - \rho_{\rm G}, \tag{4.48}$$

$$\langle \rho \rangle = \frac{\rho_{\rm G} + \rho_{\rm L}}{2}. \tag{4.49}$$

We also require the curvature of the interface $\chi(\boldsymbol{x},t)$ (positive when the center of curvature is on the liquid side), which is calculated as follows:

$$\chi(\boldsymbol{x},t) = -\nabla \cdot \boldsymbol{n}(\boldsymbol{x},t). \tag{4.50}$$

Here, $\boldsymbol{n}(\boldsymbol{x},t)$ is the unit normal vector of the interface, calculated as follows:

$$\boldsymbol{n}(\boldsymbol{x},t) = \frac{\nabla\rho(\boldsymbol{x},t)}{|\nabla\rho(\boldsymbol{x},t)|}. \tag{4.51}$$

To avoid division by zero in the above equation, we apply the approximation described for Eq. (4.25). The force $\boldsymbol{F}_{\rm sv}(\boldsymbol{x},t)$ given by Eq. (4.47) expresses the body force acting on an interface of width h, which gives the interfacial tension σ in the limit of $h \to 0$. The spatial derivatives in the above equations can be calculated by Eqs. (4.19) and (4.20). Although the last term in Eq. (4.31) also expresses the interfacial tension, the CSF model achieves higher numerical accuracy when the curvature of the interface is smaller [74].

(2) Asymptotic analysis (S-expansion)

The governing equations in the above computational scheme can be obtained through the following asymptotic analysis (S-expansion). Expanding Eqs. (4.39) and (4.43) as a Taylor series around $(\boldsymbol{x},t)$, substituting the expanded forms[5] $u_\alpha = (\Delta x)u_\alpha^{(1)} + (\Delta x)^2 u_\alpha^{(2)} + \cdots$, $p = 1 + (\Delta x)p^{(1)} + (\Delta x)^2 p^{(2)} + (\Delta x)^3 p^{(3)} + \cdots$, and $F_{{\rm sv}\alpha} = (\Delta x)^2 F_{{\rm sv}\alpha}^{(2)}$ into the Taylor expanded equations, and collecting terms with the same order of Δx, we obtain the following expressions on the diffusive time scale ($\Delta t = {\rm Sh}\Delta x = O((\Delta x)^2)$):

$$\frac{\partial u_\alpha^{(1)}}{\partial x_\alpha} = 0, \tag{4.52}$$

$$\frac{{\rm Sh}}{\Delta x}\frac{\partial u_\alpha^{(1)}}{\partial t} + u_\beta^{(1)}\frac{\partial u_\alpha^{(1)}}{\partial x_\beta} = -\frac{1}{\rho}\frac{\partial p^{(2)}}{\partial x_\alpha}$$

$$+ \frac{1}{\rho}\frac{\partial}{\partial x_\beta}\left[\frac{\mu}{\Delta x}\left(\frac{\partial u_\beta^{(1)}}{\partial x_\alpha} + \frac{\partial u_\alpha^{(1)}}{\partial x_\beta}\right)\right] + \frac{1}{\rho}F_{{\rm sv}\alpha}^{(2)}. \tag{4.53}$$

---

[5]As the density ρ is assumed to be known, it is not expanded as a power series in Δx.

Note that $p^{(1)}$ is constant on the diffusive time scale. Under appropriate initial and boundary conditions, $u_\alpha^{(2)} = p^{(3)} = 0$ in the domain. Therefore, $u_\alpha = (\Delta x)u_\alpha^{(1)} + O((\Delta x)^3)$ and $p = 1 + (\Delta x)^2 p^{(2)} + O((\Delta x)^4)$ satisfy the continuity equation (4.52) and the Navier–Stokes equations (4.53) for two-phase incompressible viscous fluids with relative errors $O((\Delta x)^2)$.

By the same procedure, the following equations are obtained on the acoustic time scale ($\Delta t = \Delta x$):

$$\frac{\partial p^{(1)}}{\partial t} + \frac{n\omega}{3}\frac{\partial u_\alpha^{(1)}}{\partial x_\alpha} = 0, \tag{4.54}$$

$$\frac{\partial u_\alpha^{(1)}}{\partial t} + \frac{1}{\rho}\frac{\partial p^{(1)}}{\partial x_\alpha} = 0. \tag{4.55}$$

Eliminating the term $u_\alpha^{(1)}$ from both equations, we get

$$\frac{\partial^2 p^{(1)}}{\partial t^2} = \frac{n}{3}\frac{\partial^2 p^{(1)}}{\partial x_\alpha^2}, \tag{4.56}$$

in the gas phase and

$$\frac{\partial^2 p^{(1)}}{\partial t^2} = \frac{n\omega_{\max}}{3\rho_{\mathrm{L}}}\frac{\partial^2 p^{(1)}}{\partial x_\alpha^2}, \tag{4.57}$$

in the liquid phase. Thus, the speed of sound becomes $c_{\mathrm{s}} = \sqrt{n/3}$ in the gas phase and $c_{\mathrm{s}} = \sqrt{n\omega_{\max}/(3\rho_{\mathrm{L}})}$ in the liquid phase. Setting $n\omega_{\max} \approx \rho_{\mathrm{L}}$, we obtain $c_{\mathrm{s}} \approx \sqrt{1/3}$ in the liquid phase, even in cases of high density ratio ($\rho_{\mathrm{L}} \gg 1$).

4.8 Computational Algorithm of Two-Phase LBM for Large Density Ratios

The computational scheme described in Secs. 4.3 and 4.7 is summarized below.

> **Step 0.** Give the initial values $\phi(\boldsymbol{x},0)$, $p(\boldsymbol{x},0)$, and $\boldsymbol{u}(\boldsymbol{x},0)$.
>
> **Step 1.** Determine $\phi(\boldsymbol{x}_{\mathrm{w}},t)$, $p(\boldsymbol{x}_{\mathrm{w}},t)$, and $\boldsymbol{u}(\boldsymbol{x}_{\mathrm{w}},t)$ on the boundary based on the given boundary conditions.
>
> **Step 2.** Compute $\phi(\boldsymbol{x},t+\Delta t)$ using Eq. (4.13) or Eq. (4.21). Also compute $\rho(\boldsymbol{x},t+\Delta t)$, $\mu(\boldsymbol{x},t+\Delta t)$, and $\omega(\boldsymbol{x},t+\Delta t)$ using Eq. (4.35) or (4.36), Eq. (4.37) or (4.38), and Eq. (4.41), respectively.

Step 3. Iterate the pressure computation $p(\boldsymbol{x}, t + \Delta t)$ as follows.

    (3-1) Set the initial value of the iteration to $p_0(\boldsymbol{x}) = p(\boldsymbol{x}, t)$.

    (3-2) Compute $p_l(\boldsymbol{x})$ by the iteration of Eq. (4.39). Repeat the iteration n times.

    (3-3) Update the pressure in the new time step as $p(\boldsymbol{x}, t + \Delta t) = p_n(\boldsymbol{x})$.

Step 4. Compute the flow velocity $\boldsymbol{u}(\boldsymbol{x}, t + \Delta t)$ at the new time step using Eq. (4.43).

Step 5. Advance the time by Δt and return to **Step 1.**

Note that in **Step 3**, $n = 2$ is sufficient even at large density ratios $(\rho_{\mathrm{L}}/\rho_{\mathrm{G}} = 1000)$, so n is not large.

4.9 Two-Phase LBM for Large Density Ratios (Method in 2004)

Although two-phase flows with large density ratios can be computed by the two-phase LBM described in Sec. 4.7, many studies have adopted the method proposed by Inamuro et al. in 2004 [69]. As this method is regarded as a milestone in two-phase LBMs for large density ratios, it is briefly presented here.

The idea of this method is to compute the velocity and pressure fields by formulating them in the LBM framework using the projection method [18]. The density ρ and viscosity coefficient μ are obtained as described in Sec. 4.7. First, the predicted flow velocity $\boldsymbol{u}^*$ is computed as follows:

$$\boldsymbol{u}^*(\boldsymbol{x}, t + \Delta t) = \sum_{i=1}^{15} c_i g_i^{\mathrm{eq}}(\boldsymbol{x} - c_i \Delta x, t), \tag{4.58}$$

where the local equilibrium distribution function g_i^{eq} is given by

$$
\begin{aligned}
g_i^{\mathrm{eq}} = E_i \bigg\{ & 1 + 3c_{i\alpha}u_\alpha + \frac{9}{2}c_{i\alpha}c_{i\beta}u_\alpha u_\beta - \frac{3}{2}u_\alpha u_\alpha \\
& + \frac{3}{4}\Delta x \left(\frac{\partial u_\beta}{\partial x_\alpha} + \frac{\partial u_\alpha}{\partial x_\beta} \right) c_{i\alpha}c_{i\beta} + 3c_{i\alpha}\frac{1}{\rho}\frac{\partial}{\partial x_\beta}\left[\mu\left(\frac{\partial u_\beta}{\partial x_\alpha} + \frac{\partial u_\alpha}{\partial x_\beta}\right)\right]\Delta x \bigg\} \\
& + E_i \frac{\kappa_g}{\rho} G_{\alpha\beta}^\rho c_{i\alpha}c_{i\beta} - \frac{2}{3}F_i \frac{\kappa_g}{\rho}|\nabla\rho|^2. \tag{4.59}
\end{aligned}
$$

The parameter $\kappa_g = O((\Delta x)^2)$ determines the interfacial tension as discussed above, F_i is a constant defined by Eq. (4.15), and $G_{\alpha\beta}^\rho$ is defined by replacing ϕ with ρ in Eq. (4.17). The interfacial tension can also be calculated by replacing ϕ with ρ in Eq. (4.33).

However, as g_i^{eq} in Eq. (4.59) contains no pressure term p, u^* is not generally divergence free (that is, $\nabla \cdot u^* \neq 0$). Thus, p is determined to satisfy $\nabla \cdot u = 0$, and u is then computed as follows:

$$\text{Sh}\frac{u - u^*}{\Delta t} = -\frac{\nabla p}{\rho}, \tag{4.60}$$

where p is computed by solving the following Poisson equation:

$$\nabla \cdot \left(\frac{\nabla p}{\rho}\right) = \text{Sh}\frac{\nabla \cdot u^*}{\Delta t}. \tag{4.61}$$

The above Poisson equation can be solved by various methods. Here we adopt the LBM. That is, we introduce a new velocity distribution function h_i and iterate it as follows:

$$h_i^{n+1}(x + c_i\Delta x) = h_i^n(x) - \frac{1}{\tau_h}[h_i^n(x) - E_i p^n(x)] - \frac{1}{3}E_i\frac{\partial u_\alpha^*}{\partial x_\alpha}\Delta x, \tag{4.62}$$

where n is the number of iterations and the relaxation time τ_h is given by

$$\tau_h = \frac{1}{\rho} + \frac{1}{2}. \tag{4.63}$$

The pressure p is obtained as follows:

$$p = \sum_{i=1}^{15} h_i. \tag{4.64}$$

Equation (4.62) is iterated until $|p^{n+1} - p^n|/\rho < \varepsilon_0$ (e.g., $\varepsilon_0 = 10^{-6}$), and p is then obtained.

This method performs stable computations up to a density ratio of 1000 [70], but requires more iterations of Eq. (4.62) as the density ratio increases. The computational time at a density ratio of 50 was approximately 50 times longer than that of the method in Sec. 4.7 [74].

4.10   Numerical Examples

In the following numerical examples, the Cahn–Hilliard equation is used as the free energy model. Unless otherwise specified, the mobility is set to $M_\phi = 0.5$.

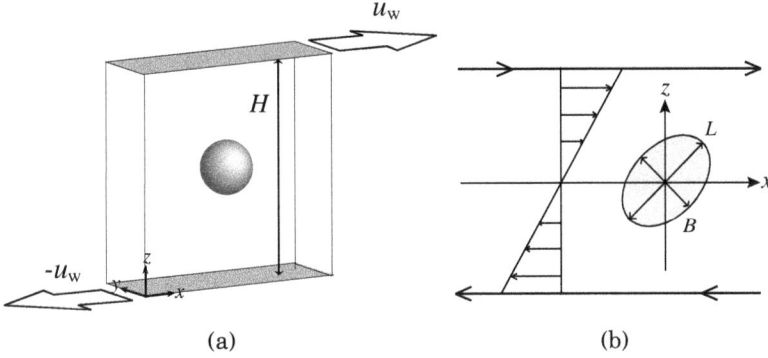

Fig. 4.2 (a) Droplet in a shear flow; (b) deformation degree $D_d = (L - B)/(L + B)$.

4.10.1 *Droplet deformation and breakup in shear flows*

We first show the deformation and breakup of a droplet in shear flows. These results were obtained by the equal-density two-phase LBM developed in Sec. 4.5 [68]. A droplet of radius R is placed in the center between two parallel plates of height H (Fig. 4.2(a)). Starting from $t = 0$, the upper and lower walls are moved in opposite directions at a speed of u_w, and the deformation and breaking states of the droplet are tracked. The no-slip (counter-slip) boundary condition is imposed on the upper and lower walls, and the periodic boundary condition is imposed on the side surfaces. The computational domain is divided into a $128\Delta x \times 64\Delta x \times 128\Delta x$ cubic lattice. The dimensionless parameters of this problem are the Reynolds number $\mathrm{Re} = \rho \Gamma R^2/\mu_c$, the capillary number $\mathrm{Ca} = \mu_c \Gamma R/\sigma$, and the viscosity ratio $\eta = \mu_d/\mu_c$, where $\Gamma = 2u_w/H$ is the shear strength, μ_c is the viscosity coefficient of the continuous phase, and μ_d is the viscosity coefficient of the droplet. The computational conditions are $a = 9/49$, $b = 2/21$, $T = 0.55$ (in this case, $\phi_2 = 4.895$ and $\phi_1 = 2.211$),[6] $\kappa_f = 0.01(\Delta x)^2$, and $R = 16\Delta x$. The other values of u_w, τ_g, and κ_g are varied to change the Reynolds number as $\mathrm{Re} = 0.20$ and 10 and the capillary number as $0.1 \leq \mathrm{Ca} \leq 0.45$. The viscosity ratio is $\eta = 1$. The deformation degree D_d of the droplet is defined as $D_d = (L - B)/(L + B)$, where L and B are the lengths of the major and minor axes of the droplet, respectively (see Fig. 4.2(b)).

Figure 4.3 shows the deformation results of the droplet in steady state.

---

[6]Note that the actual ϕ_{max} and ϕ_{min} in the computation deviate slightly from ϕ_2 and ϕ_1, respectively.

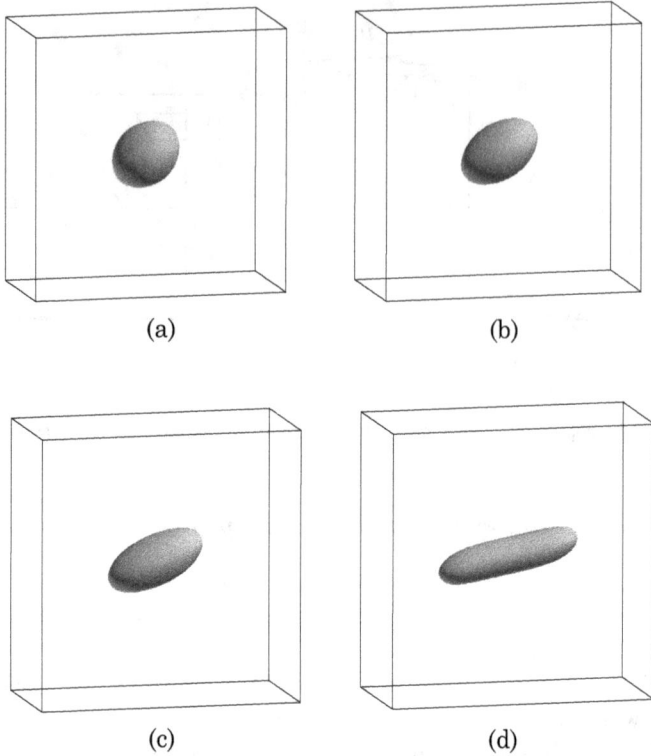

Fig. 4.3 Deformation of a droplet (Re = 0.20) in flows with different capillary numbers: (a) Ca = 0.10; (b) Ca = 0.20; (c) Ca = 0.30; (d) Ca = 0.40. Reprinted from Fig. 3 in [68] with permission from World Scientific Publishing.

Here, Re is fixed at 0.20 and Ca is varied as 0.10, 0.20, 0.30, and 0.40. The droplet transforms into an ellipsoid, becoming longer and thinner with increasing Ca. In steady state, the shear force is considered to balance the interfacial tension. Figure 4.4 shows the result of slightly increasing Ca to 0.45 without changing the Re. In this case, the ellipsoid transformation is followed by transformation into a dumbbell shape, which finally breaks to form two droplets.

Figure 4.5 plots the relationship between capillary number Ca and deformation degree D_d of the droplet under various conditions. For comparison, we also plot the results of existing experiments [22, 116], numerical studies [92], and theoretical analysis [21]. The existing results were obtained for very low Reynolds numbers (Re < 10^{-2}) and Re = 0 in the theoretical

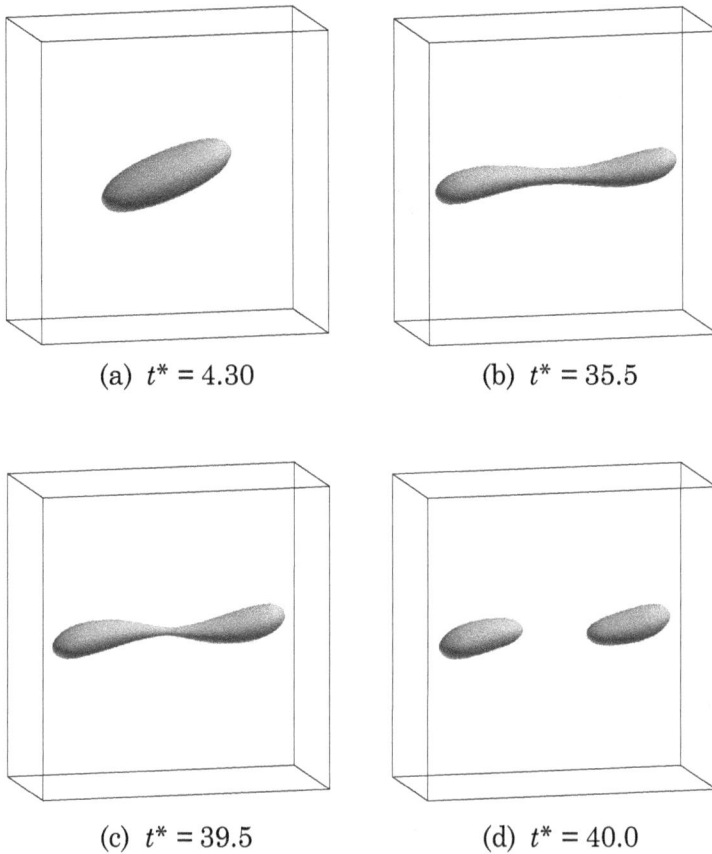

(a) $t^* = 4.30$

(b) $t^* = 35.5$

(c) $t^* = 39.5$

(d) $t^* = 40.0$

Fig. 4.4   Breakup of a droplet (Re = 0.20, Ca = 0.45, $t^* = t\Gamma/\text{Sh}$).

analysis. As shown in the figure, our results for Ca = 0.10, 0.20, 0.30 and Re = 0.20 agree very well with the existing results, but our deformation degree D_d for Ca = 0.40 is slightly larger than the existing results at low Reynolds numbers. Furthermore, when Ca = 0.45, the droplet breaks up in the present computation but does not break at low Reynolds numbers in the previous studies. These differences might be explained by the different Reynolds numbers. Indeed, when the Reynolds number increases to Re = 10, the calculated result exhibits similar break-up behavior, indicating that the droplet is more likely to deform and break up at higher Reynolds numbers.

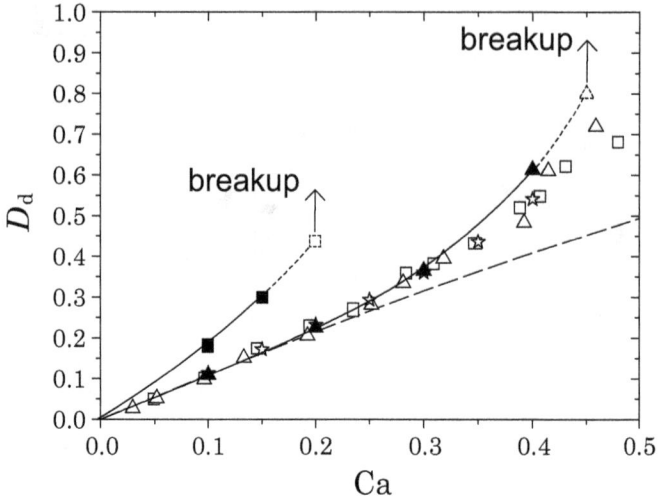

Fig. 4.5 Deformation degree D_d versus capillary number Ca: ▲Re = 0.20, ■Re = 10 (present); – – – Cox [21] (theoretical); △Rumcheidt and Mason [116] (experimental); □de Bruijn [22] (experimental); ☆Li et al. [92] (numerical).

The behavior of a droplet in a tube with a square cross section has been investigated using the same method [82].

4.10.2 *Layered Poiseuille flow*

As a simple validation of the two-phase LBM for large density ratios presented in Sec. 4.8, we consider a layered Poiseuille flow. As shown in Fig. 4.6, a gas phase of density ρ_G and viscosity coefficient μ_G is placed in the region $|y| < D/2$ between two stationary walls at $y = D$ and $y = -D$, and a liquid phase of density ρ_L and viscosity coefficient μ_L is placed in the region $D/2 \le |y| \le D$. Two-phase LBMs are commonly validated in layered Poiseuille flows, but in most studies, the flow is induced by a body force (such as gravity). In such cases, the pressure computation cannot be verified because the pressure in the region becomes constant. To avoid this problem, we induce flow by applying a pressure difference Δp between the inlet ($x = 0$) and outlet ($x = L$). The numerical accuracy is validated by comparing the velocity distribution and pressure distribution with their corresponding analytical solutions. The velocity distribution $u(y)$ in the x-direction is analytically calculated as follows:

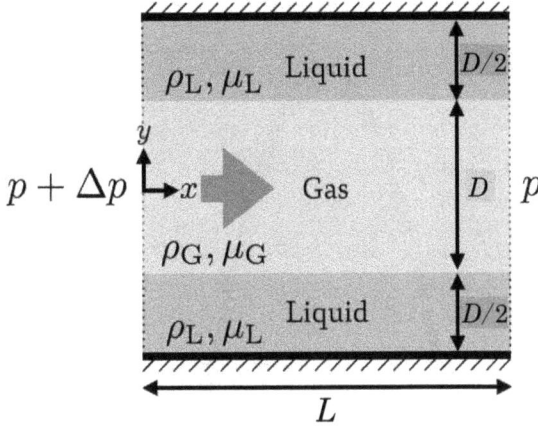

Fig. 4.6   Computational domain of a layered Poiseuille flow.

$$u_{\text{th}}(y) = \begin{cases} \frac{\Delta p}{L}\left[-\frac{1}{2\mu_G}y^2 + \frac{1}{8}\left(\frac{1}{\mu_G} + \frac{3}{\mu_L}\right)D^2\right], & |y| \leq D/2, \\ \frac{\Delta p}{L}\left(-\frac{1}{2\mu_L}y^2 + \frac{1}{2\mu_L}D^2\right), & D/2 \leq |y| \leq D. \end{cases} \tag{4.65}$$

The computational domain is divided into a $25\Delta x \times 400\Delta x \times 4\Delta x$ cubic lattice. The specular condition is used in the z-direction. Here we show the calculated results of two cases with different density ratios: $\rho_L/\rho_G = 50$ ($\rho_G = 1$, $\mu_G = 0.032\Delta x$; $\rho_L = 50$, $\mu_L = 1.6\Delta x$) and $\rho_L/\rho_G = 1000$ ($\rho_G = 1$, $\mu_G = 0.032\Delta x$; $\rho_L = 1000$, $\mu_L = 32\Delta x$). The pressure differences in the $\rho_L/\rho_G = 50$ and 1000 cases are determined as $\Delta p = 2.415 \times 10^{-7}$ and 2.550×10^{-7}, respectively, giving a Reynolds number of Re $= \rho_G u_{\text{th}}(0)D/\mu_G = 10$ in both cases. The parameters in Eq. (4.16) are $a = 1$, $b = 1$, and $T = 2.93 \times 10^{-1}$ (in this case, $\phi_2 = 4.031 \times 10^{-1}$ and $\phi_1 = 2.638 \times 10^{-1}$). In the $\rho_L/\rho_G = 50$ case, Eq. (4.39) is iterated twice and ω_{max} in Eq. (4.41) is set to 25. In the $\rho_L/\rho_G = 1000$ case, Eq. (4.39) is iterated twice and ω_{max} is set to 500. The other parameters are $\kappa_f = 0.06(\Delta x)^2$, $\lambda = 0$, and $\sigma = 0$. The viscosity coefficient μ at the interface is calculated by linear interpolation of Eq. (4.37) and harmonic interpolation of Eq. (4.38), and the profiles of the resulting flow velocities u in the x-direction are compared in Fig. 4.7. The calculation remains stable at a density ratio of 1000. Furthermore, at both density ratios, the harmonic interpolation of the interfacial viscosity coefficient gives similar results to the analytical solution.

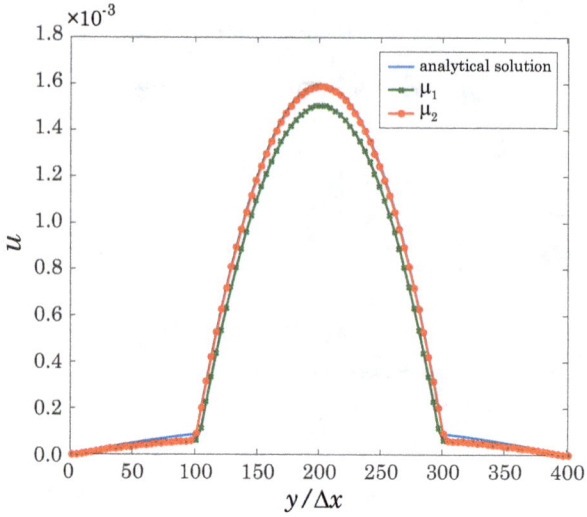

(a)

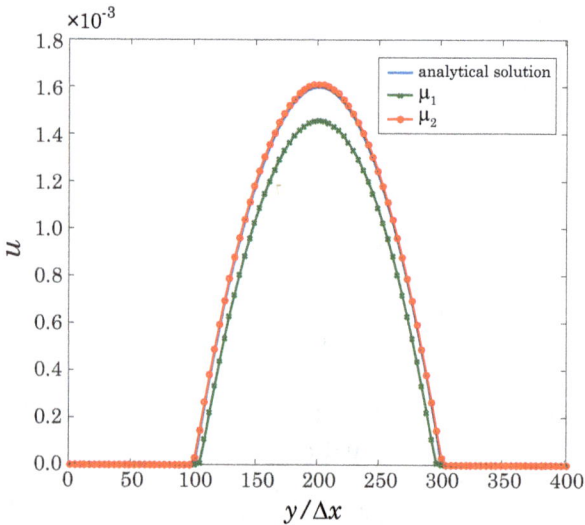

(b)

Fig. 4.7   Calculated velocity profiles: (a) $\rho_L/\rho_G = 50$; (b) $\rho_L/\rho_G = 1000$. The viscosity coefficients at the interface were calculated by linear interpolation of Eq. (4.37) (μ_1) and harmonic interpolation of Eq. (4.38) (μ_2). Reprinted from Figs. 2(b) and 3(b) in [76] with permission from Elsevier.

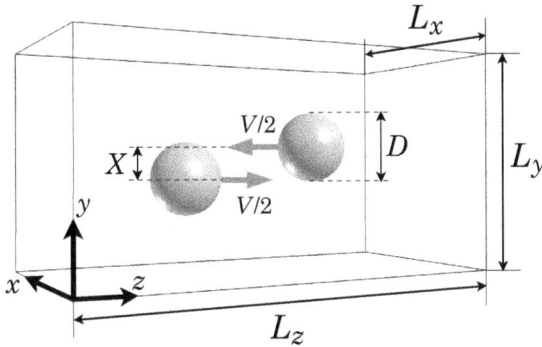

Fig. 4.8   Computational domain of droplet collisions.

4.10.3   *Droplet collision*

Collision dynamics of binary droplets are involved in many natural and artificial phenomena, such as raindrop formation, spray coating, ink-jet printing, and processes in dispersed phases. Therefore, droplet collision dynamics have been extensively studied in experiments, theoretical analyses, and numerical simulations. Collision outcomes depend on the Weber number We and the impact parameter B (defined later) and are classified into five groups [110]: coalescence after minor deformation at very small We (Group I), bouncing at small We (Group II), coalescence at moderate We and B (Group III), reflexive separation at moderate We and small B (Group IV), and stretching separation at moderate We and large B (Group V). Recently, Pan et al. [105] experimentally demonstrated a new collision outcome called rotational separation, which belongs to a small subgroup of Group III. Rotational separation, which occurs when colliding droplets rotate by more than 180 degrees, differs from reflexive and stretching separations. As rotational separation is limited to a narrow region of the We–B diagram, it is suitable for validating the two-phase LBM for large density ratios described in Sec. 4.8.

We first simulate head-on binary droplet collisions at various Weber and Reynolds numbers. The computational domain is divided into a $192\Delta x \times 96\Delta x \times 96\Delta x$ cubic lattice (Fig. 4.8). Two liquid droplets with the same diameter ($D = 32\Delta x$) are placed $2D$ apart in a gas and accelerated to a relative velocity of V. Periodic boundary conditions are imposed on all sides of the domain. Gravitational effects are ignored in droplet collision

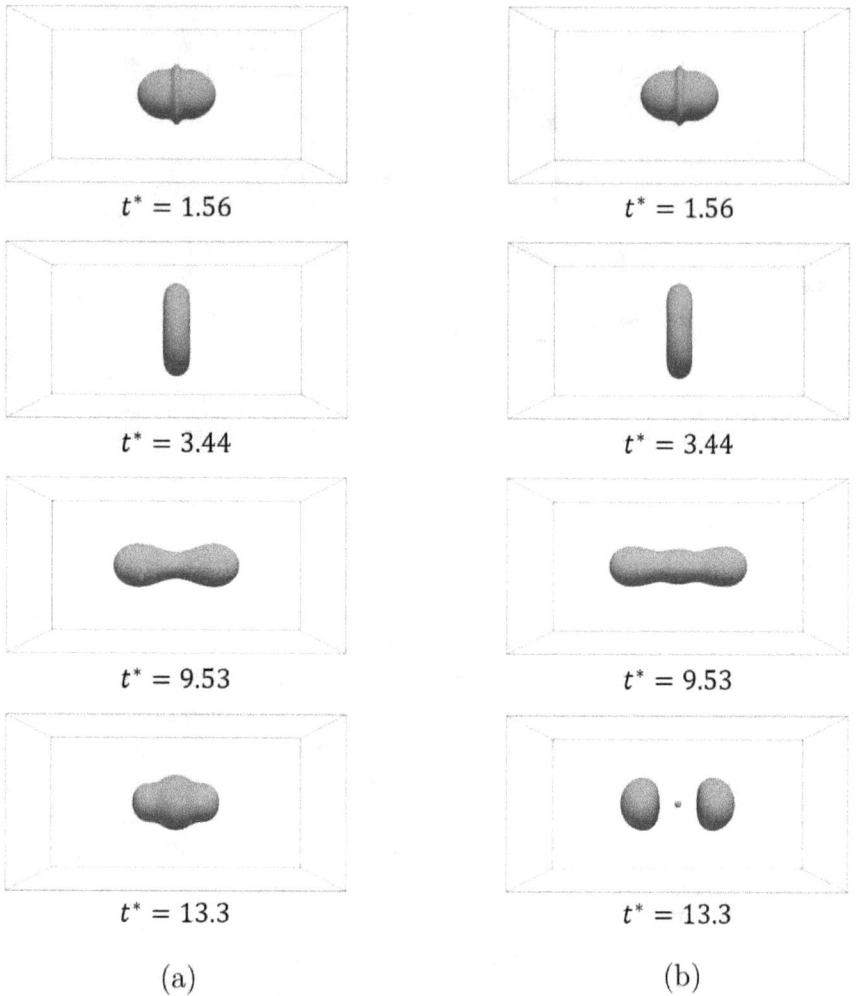

Fig. 4.9 Time variation of the droplet shape in a head-on collision in two cases: (a) We = 30 and Re = 1000; (b) We = 35 and Re = 1000. The density ratio is $\rho_L/\rho_G = 50$, and $t^* = tV/(\mathrm{Sh}D)$.

problems. The liquid-to-gas density and viscosity ratios are $\rho_L/\rho_G = 50$ ($\rho_L = 50$, $\rho_G = 1$) and $\mu_L/\mu_G = 50$, respectively. The relative velocity is constant at $V = 5.0 \times 10^{-3}$, and the viscosity coefficients of the liquid and gas are changed within $2.67 \times 10^{-3} < \mu_L/\Delta x < 4.00 \times 10^{-2}$ and $5.33 \times$

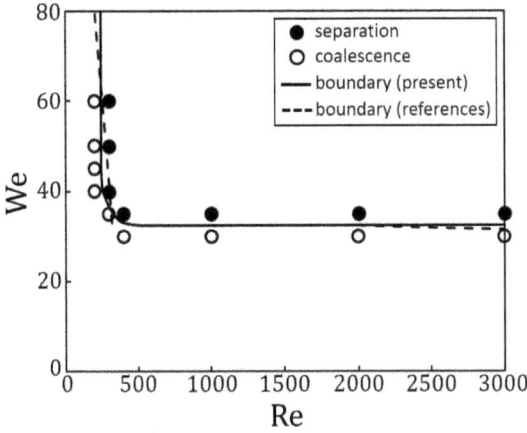

Fig. 4.10 Boundaries between coalescence and separation in the Re–We diagram. The solid and dashed lines indicate the boundaries of the present results and of the results by the VOF method [36, 118], respectively.

$10^{-5} < \mu_G/\Delta x < 8.00 \times 10^{-4}$, respectively, maintaining a constant viscosity ratio ($\mu_L/\mu_G = 50$). Meanwhile, the interfacial tension is changed within $6.67 \times 10^{-4} < \sigma/\Delta x < 1.33 \times 10^{-3}$. The dimensionless parameters of this problem are the Reynolds number $\text{Re} = \rho_L D V/\mu_L$, the Ohnesorge number $\text{Oh} = \mu_L/\sqrt{\rho_L D \sigma}$, and the Weber number $\text{We} = \rho_L D V^2/\sigma$. Only two of these three parameters are independent. Here we select the Reynolds and Weber numbers and set $a = 1$, $b = 1$, and $T = 2.93 \times 10^{-1}$. It follows that the maximum and minimum values of the order parameter are $\phi_2 = 4.031 \times 10^{-1}$ and $\phi_1 = 2.638 \times 10^{-1}$, respectively. The mobility is set to $M_\phi = 0.01\Delta x$. The other parameters, namely, κ_f, n in Eq. (4.39), ω_{max} in Eq. (4.41), and λ in Eq. (4.43), are set to $\kappa_f = 0.06(\Delta x)^2$, $n = 2$, $\omega_{max} = 12.5$, and $\lambda = 1$, respectively. The viscosity coefficient in the interface is linearly interpolated by Eq. (4.37). In the simulations, the interfacial tension is increased linearly in the range $1000\Delta t \leq t \leq 2000\Delta t$ and the calculation is subsequently continued until $t = 5000\Delta t$. At this time, the droplets are stationary in the equilibrium state and the time is reset to $t = 0$. The velocities of the two droplets are then changed linearly from 0 to their prescribed values $\pm V/2$ during $100\Delta t$.

Figure 4.9 shows the temporal changes in the calculated droplet shapes in two representative cases ($\text{We} = 30$ and 35 with $\text{Re} = 1000$). In the first case, the droplets collide head on and form a disk-like shape, which evolves into a dumbbell shape (Fig. 4.9(a)). The droplet then oscillates repeatedly and finally transforms into a single spherical droplet. This case is characterized by coalescence. The second case (Fig. 4.9(b)) displays similar behavior up to $t^* = 3.44$. Thereafter, a long cylinder with rounded ends is formed. At $t^* = 9.53$, the shapes developed in the two cases are quite different. At $t^* = 13.3$, the cylindrical droplet breaks into two major droplets and a smaller satellite droplet in the middle. This case is characterized by separation.

Performing similar simulations at other Weber and Reynolds numbers, we can investigate the critical Weber number between coalescence and separation. Figure 4.10 shows the coalescence–separation boundaries in a Re–We diagram. In the present study, we change the Weber number in intervals of 5 by changing σ, maintaining a fixed Reynolds number. The critical Weber number is then identified as an intermediate value between coalescence and separation. In this figure, the solid line represents the boundary of the present results, and the dashed lines for Re < 400 and Re > 2000 indicate the boundaries of the results by Finotello et al. [36] and Saroka et al. [118], respectively. Both reference boundaries were obtained by the VOF method. In the Re < 500 region, the droplets easily coalesce and the critical Weber number significantly increases with decreasing Reynolds number. Conversely, in the Re > 1000 region, the critical Weber number is almost independent of the Reynolds number. Our critical Weber numbers are consistent with the results of the VOF method at various Reynolds numbers. Binary head-on collisions of unequally sized droplets have also been simulated using the method described in Sec. 4.9 [163].

To confirm the development of rotational separation, we now perform an offset collision calculation [142]. The computational domain is divided into a $192\Delta x \times 192\Delta x \times 384\Delta x$ cubic lattice (Fig. 4.8). Two droplets of diameter $D = 64\Delta x$, initially separated by $2D$, collide at a relative velocity of V. The length X in Fig. 4.8 is the distance between the center of one droplet and the relative velocity vector passing through the other droplet center. Periodic boundary conditions are imposed on all outer boundaries. The liquid-to-gas ratios of density and viscosity are $\rho_L/\rho_G = 600$ ($\rho_L = 600$, $\rho_G = 1$) and $\mu_L/\mu_G = 70$ ($\mu_L = 4.32000 \times 10^{-1}\Delta x$, $\mu_G = 6.17143 \times 10^{-3}\Delta x$), respectively. The same conditions were prepared in Pan et al.'s experiment [105] of dodecane droplets in air. The dimensionless parameters of this problem are the impact parameter $B = X/D$ and the above-mentioned Ohnesorge and Weber numbers. To maintain the experimental value of Oh (= 0.0126), we set $\sigma = 3.05907 \times 10^{-2}\Delta x$, and vary V and X such that

$30 \leq We \leq 40$ and $0.30 \leq B \leq 0.50$, respectively. For example, to obtain $We = 35.4$, we set $V = 5.31044 \times 10^{-3}$. The parameters in Eq. (4.16) are $a = 1$, $b = 1$, and $T = 2.93 \times 10^{-1}$ (in this case, $\phi_2 = 4.031 \times 10^{-1}$ and $\phi_1 = 2.638 \times 10^{-1}$). We also set $\lambda = 10$ in Eq. (4.43), $\kappa_f = 0.04(\Delta x)^2$, $n = 2$ in Eq. (4.39), and $\omega_{max} = 300$ in Eq. (4.41). The viscosity coefficient at the interface is obtained by linear interpolation of Eq. (4.37).

Figure 4.11 compares the calculated and experimental changes in droplet shapes for $We = 35.4$, $B = 0.42$, and $Oh = 0.0126$. The droplet shape undergoes complicated changes due to the combination of rotational motion, flattening motion (pancake shape), and extension motion (tube shape or dumbbell shape). These motions are caused by combinations of the outer rotating flow and the inner reflecting flow in the coalescing droplet. The rotational motion persists but the flattening motion ($t = 0.98T$) gradually shifts to the extension motion ($t = 1.98T$). Finally the droplet transforms into a dumbbell shape ($t = 2.78T$) and separates. The temporal changes in the calculated droplet shape agree with the experimentally observed changes.

Coalescence (Group III), reflection separation (Group IV), and stretching separation (Group V) have also been investigated using the method in Sec. 4.9 [71, 117].

4.10.4 *Milk crown*

A milk crown occurs when a droplet collides with a thin liquid film. Simulating this difficult problem has long been a dream of computational science. The milk crown phenomenon is characterized by large density ratio, generation of a very thin liquid sheet, and formation of a crown due to instability of the liquid sheet, all of which are difficult to simulate. Therefore, the milk crown problem is a suitable validation problem for the two-phase LBM with large density ratios described in Sec. 4.8.

Figure 4.12 shows the computational domain of this problem. A droplet of diameter $D = 240\Delta x$ collides with a liquid film of thickness $H = D/8 = 30\Delta x$. Initially, the droplet and liquid film are separated by $G = D/12 = 20\Delta x$, and the initial velocity of the droplet is set to $U = 0.0125$. The gravitational effect is embodied in the term $-3E_i c_{iz} [1 - (\rho_G/\rho)] g\Delta x$ (where $g = O((\Delta x)^2)$ is the gravitational acceleration), which is added to the right-hand side of Eq. (4.42). The computational domain is divided into a $1680\Delta x \times 1680\Delta x \times 720\Delta x$ cubic lattice. The specular condition is imposed on all outer surfaces of the computational domain. The liquid-to-gas

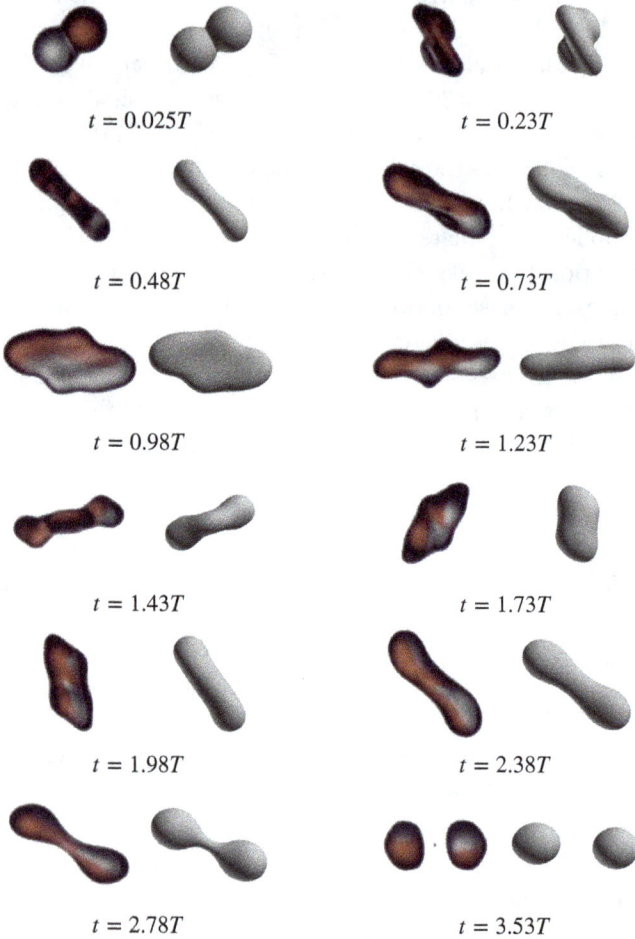

Fig. 4.11 Temporal changes in droplet shape with $B = 0.42$, We $= 35.4$, Oh $= 0.0126$, and $\rho_L/\rho_G = 600$, where $T = (\pi/4)\sqrt{\rho_L D^3/\sigma}$. The left and right at each time present the experimental results of Pan et al. [105] and the present calculated results, respectively. Reprinted from Fig. 5 in [142] with permission from AIP Publishing.

density ratio is $\rho_L/\rho_G = 800$ ($\rho_L = 800$, $\rho_G = 1$), and the viscosity coefficients are $\mu_L = 1.2\Delta x$ ($\nu_L = \mu_L/\rho_L = 1.5 \times 10^{-3}\Delta x$) and $\mu_G = 1.2 \times 10^{-2}\Delta x$ ($\nu_G = \mu_G/\rho_G = 1.2 \times 10^{-2}\Delta x$). The parameters in Eq. (4.16) are set to $a = 1$, $b = 1$, and $T = 2.93 \times 10^{-1}$ (in this case, $\phi_2 = 4.031 \times 10^{-1}$ and

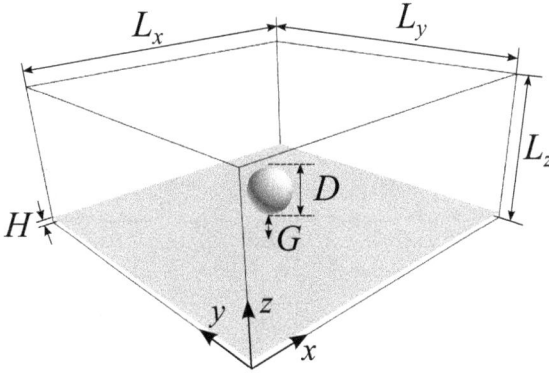

Fig. 4.12   Computational domain of a milk crown.

$\phi_1 = 2.638 \times 10^{-1}$). In Eq. (4.43), we set $\lambda = 1.0$. The governing parameters of this problem are the Reynolds number Re $= \rho_L U D / \mu_L$, the Weber number We $= \rho_L U^2 D / \sigma$, and the Froude number Fr $= U/\sqrt{gD}$. We set Re $= 2000$, We $= 250, 426, 1000$, and Fr $= 8.75$. The other parameters are set to $\kappa_f = 0.06(\Delta x)^2$, $\sigma = 1.2 \times 10^{-1} \Delta x$ for We $= 250$, $\sigma = 7.04 \times 10^{-2} \Delta x$ for We $= 426$, and $\sigma = 3.0 \times 10^{-2} \Delta x$ for We $= 1000$. In Eqs. (4.39) and (4.41), we set $n = 2$ and $\omega_{max} = 200$, respectively. The viscosity coefficient at the interface is obtained by linear interpolation of Eq. (4.37). The vertical velocity of the liquid film surface is initially disturbed by a Gaussian distribution with a standard deviation of $0.75U$. This disturbance is large but rapidly decays due to surface tension of the liquid film.

Figure 4.13 shows the calculated results for We $= 426$. The initial disturbance on the liquid film surface remains until $t^* = 0.05$. Immediately after the droplet collision, a thin liquid sheet (crown sheet) grows in the circumferential and vertical directions. As the liquid sheet moves outward, the symmetry in the circumferential direction is broken, and small droplets finally scatter from the sheet tip. Figure 4.14 shows the density distribution on $y = L_y/2$. This figure clarifies the spread of the liquid sheet. Figure 4.15 compares the temporal changes in the liquid sheet radius in our calculations with the theoretical predictions by Yarin and Weiss [159] and Rieber and Frohn [113]. Exploiting symmetry of the $y = L_y/2$ and $x = L_x/2$ planes, the simulation is performed in a quarter domain to reduce the computational time. As clarified in the figure, the time variation of the crown sheet radius is independent of the Weber number and the calculated results agree with the theoretical predictions.

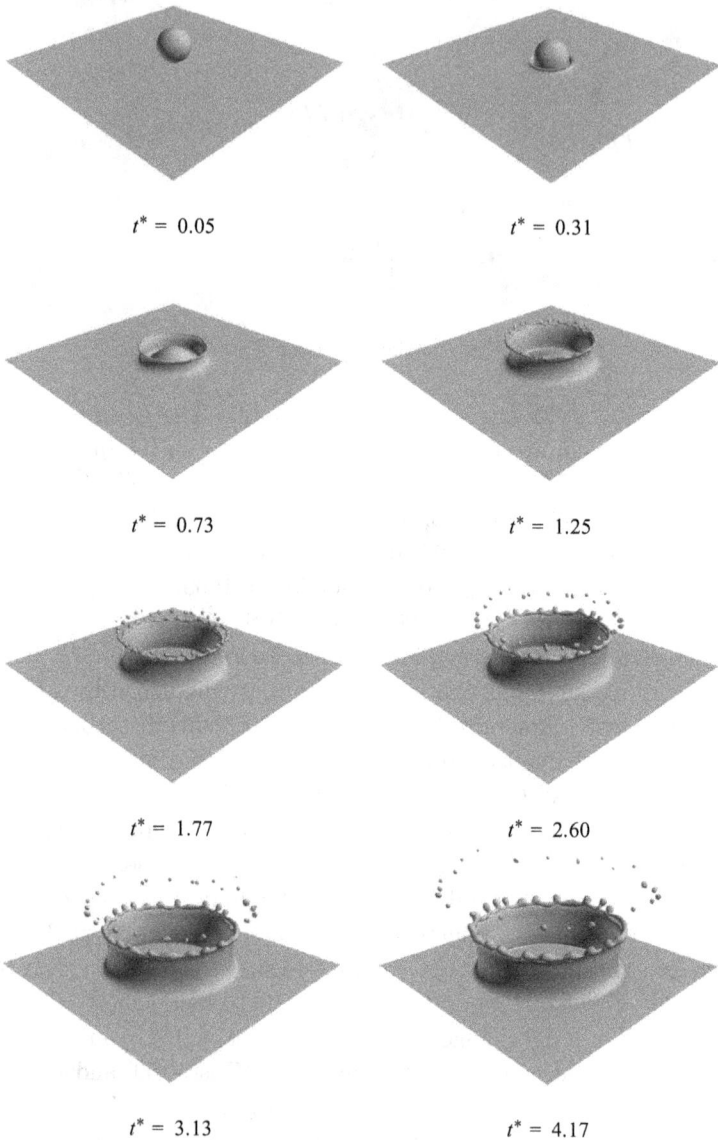

Fig. 4.13 Calculated results of a milk crown with We = 426, Re = 2000, Fr = 8.75, and $\rho_L/\rho_G = 800$, where $t^* = tU/(\text{Sh}D)$. Reprinted from Fig. 6 in [76] with permission from Elsevier.

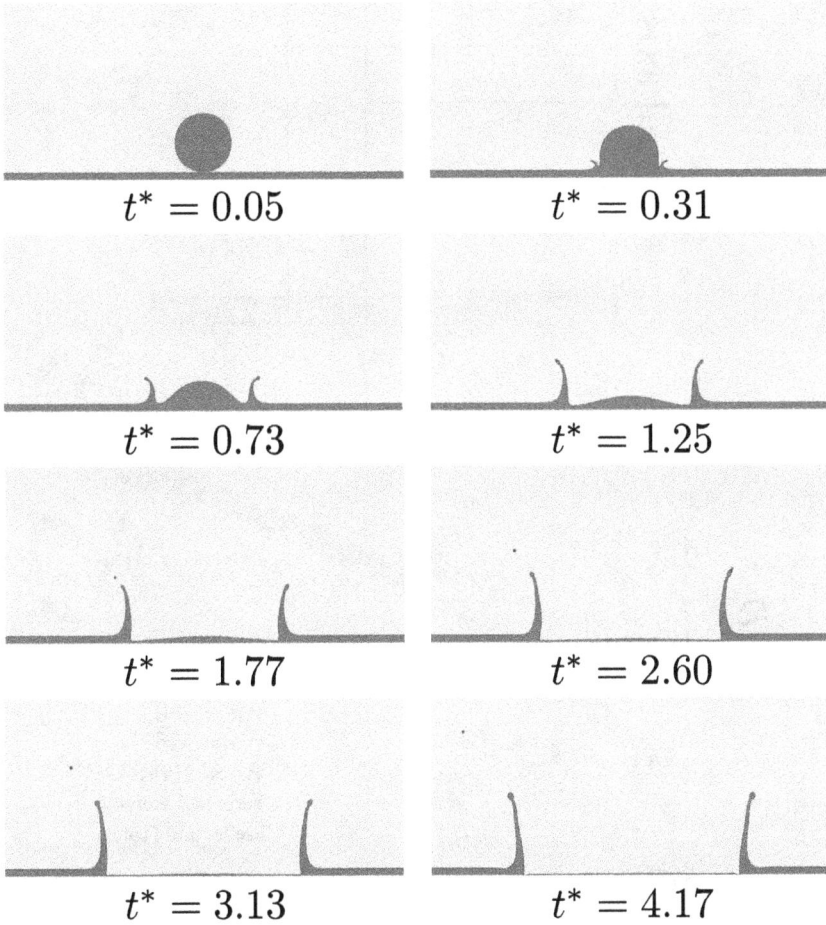

Fig. 4.14 Density distribution on $y = L_y/2$ with We = 426, Re = 2000, Fr = 8.75, and ρ_L/ρ_G = 800, where $t^* = tU/(\text{Sh}D)$. Reprinted from Fig. 9 in [76] with permission from Elsevier.

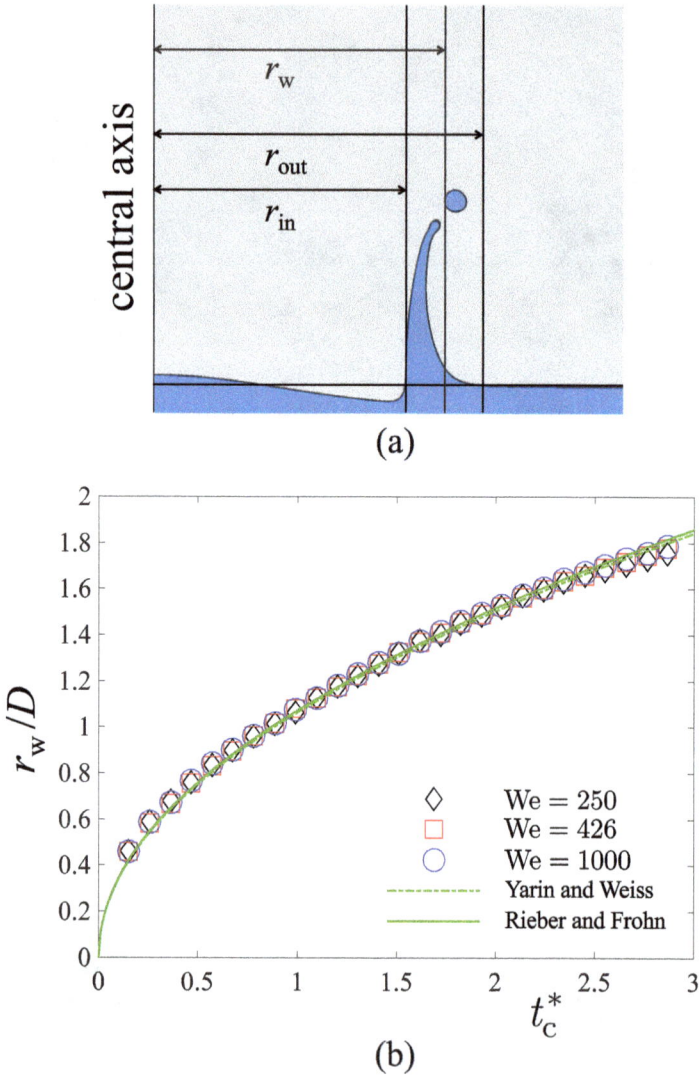

Fig. 4.15   Temporal changes in a liquid sheet radius with We $= 250, 426, 1000$, Re $= 2000$, Fr $= 8.75$, and $\rho_\mathrm{L}/\rho_\mathrm{G} = 800$, where $t_\mathrm{c}^* = t_\mathrm{c} U/(\mathrm{Sh}D)$: (a) radius of the liquid sheet $r_\mathrm{w} = (r_\mathrm{in} + r_\mathrm{out})/2$; (b) comparison with the theoretical predictions by Yarin and Weiss [159] and Rieber and Frohn [113]. The time is the elapsed time from impact t_c. Reprinted from Fig. 11 in [76] with permission from Elsevier.

Appendix A

Definitions of Dimensionless Variables

We first define the main dimensionless variables used in Chaps. 1 to 3. For this purpose, we require a characteristic length $\hat{L}$, a characteristic particle speed $\hat{c}$, a characteristic time $\hat{t}_0$, a reference density (fluid density) $\hat{\rho}_0$, a reference temperature $\hat{T}_0$, and a specific heat at constant pressure $\hat{c}_p$. We define two characteristic times: a diffusive time scale $\hat{t}_0 = \hat{L}/\hat{U}$ (where $\hat{U}$ is a characteristic flow speed) for handling slow phenomena and an acoustic time scale $\hat{t}_0 = \hat{L}/\hat{c}$ for handling fast phenomena. The symbol $\hat{\ }$ denotes a dimensional variable.

$$
\left\{
\begin{aligned}
&c_i = \hat{c}_i/\hat{c}, & &x = \hat{x}/\hat{L}, & &t = \hat{t}/\hat{t}_0, \\
&f_i = \hat{f}_i/\hat{\rho}_0, & &g_i = \hat{g}_i/\hat{T}_0, & & \\
&\rho = \hat{\rho}/\hat{\rho}_0, & &p = \hat{p}/(\hat{\rho}_0\hat{c}^2), & & \\
&u = \hat{u}/\hat{c}, & &u_T = \hat{u}_T/\hat{c}, & & \\
&e = \hat{e}/\hat{c}^2, & &T = \hat{T}/\hat{T}_0, & &q = \hat{q}/(\hat{\rho}_0\hat{c}_p\hat{c}\hat{T}_0), \\
&\nu = \hat{\nu}/(\hat{c}\hat{L}), & &\zeta = \hat{\zeta}/(\hat{c}\hat{L}), & & \\
&\alpha = \hat{\alpha}/(\hat{c}\hat{L}), & &k = \hat{k}/(\hat{\rho}_0\hat{c}_p\hat{c}\hat{L}), & & \\
&G = \hat{G}\hat{L}/\hat{c}^2 \ \ (\text{Chaps. 1 and 2}), & & & & \\
&g = \hat{g}\hat{L}/\hat{c}^2, & &\beta = \hat{\beta}\hat{T}_0, & &\omega = \hat{\omega}\hat{L}/\hat{c}, \\
&G = \hat{G}\hat{L}/(\hat{\rho}_0\hat{c}^2) \ \ (\text{Chap. 3}), & & & & \\
&X = \hat{X}/\hat{L}, & &U = \hat{U}/\hat{c}, & & \\
&F = \hat{F}/(\hat{\rho}_0\hat{c}^2\hat{L}^2), & &T = \hat{T}/(\hat{\rho}_0\hat{c}^2\hat{L}^3), & & \\
&P = \hat{P}/(\hat{\rho}_0\hat{c}), & &L = \hat{L}/(\hat{\rho}_0\hat{c}\hat{L}), & & \\
&\Omega = \hat{\Omega}\hat{L}/\hat{c}, & & & & \\
&M = \hat{M}/(\hat{\rho}_0\hat{L}^3), & &I = \hat{I}/(\hat{\rho}_0\hat{L}^5), & & \\
&Q = \hat{Q}\hat{L}/(\hat{\rho}_0\hat{c}_p\hat{T}_0\hat{c}). & & & &
\end{aligned}
\right.
\tag{A.1}
$$

We now define the main dimensionless variables used in Chap. 4. For this purpose, we require a characteristic length $\hat{L}$, a characteristic particle speed $\hat{c}$, a characteristic time $\hat{t}_0$, a reference value of order parameter $\hat{\phi}_0$, and a reference density (gas phase density) $\hat{\rho}_0$. Again we define a diffusive time scale $\hat{t}_0 = \hat{L}/\hat{U}$ (where $\hat{U}$ is a characteristic flow speed) for handling slow phenomena and an acoustic time scale $\hat{t}_0 = \hat{L}/\hat{c}$ for handling fast phenomena. The symbol $\hat{\ }$ denotes a dimensional variable.

$$\begin{cases} \boldsymbol{c}_i = \hat{\boldsymbol{c}}_i/\hat{c}, & \boldsymbol{x} = \hat{\boldsymbol{x}}/\hat{L}, & t = \hat{t}/\hat{t}_0, \\ \phi = \hat{\phi}/\hat{\phi}_0, & \rho = \hat{\rho}/\hat{\rho}_0, & g_i = \hat{g}_i/\hat{\rho}_0, \\ \boldsymbol{u} = \hat{\boldsymbol{u}}/\hat{c}, & p = \hat{p}/(\hat{\rho}_0\hat{c}^2), & h_i = \hat{h}_i/(\hat{\rho}_0\hat{c}^2), \\ \mu = \hat{\mu}/(\hat{\rho}_0\hat{c}\hat{L}), & \sigma = \hat{\sigma}/(\hat{\rho}_0\hat{c}^2\hat{L}), & M_\phi = \hat{M}_\phi/(\hat{c}\hat{L}), \\ g = \hat{g}\hat{L}/\hat{c}^2. \end{cases} \qquad \text{(A.2)}$$

Note that the time step Δt and the lattice spacing Δx in the LBM and LKS are related by $\Delta t = \text{Sh}\Delta x$ (where $\text{Sh} = \hat{L}/(\hat{t}_0\hat{c}) = \hat{U}/\hat{c} = O(\Delta x)$) on the diffusive time scale and $\Delta t = \Delta x$ on the acoustic time scale.

Appendix B

D3Q19 and D3Q27 Models

As shown in Fig. B, the particle velocities of the three-dimensional nineteen-velocity model (D3Q19 model) are given by

$$c_i = \begin{cases} (0,0,0), & i = 1, \\ (\pm 1, 0, 0), (0, \pm 1, 0), (0, 0, \pm 1), & i = 2, 3, \cdots, 7, \\ (\pm 1, \pm 1, 0), (\pm 1, 0, \pm 1), (0, \pm 1, \pm 1), & i = 8, 9, \cdots, 19. \end{cases} \quad \text{(B.1)}$$

The constants of the local equilibrium distribution function (1.17) are

$$E_i = \begin{cases} \frac{1}{3}, & i = 1, \\ \frac{1}{18}, & i = 2, 3, \cdots, 7, \\ \frac{1}{36}, & i = 8, 9, \cdots, 19. \end{cases} \quad \text{(B.2)}$$

The particle velocities in the three-dimensional twenty-seven-velocity model (D3Q27 model) are given by

$$c_i = \begin{cases} (0,0,0), & i = 1, \\ (\pm 1, 0, 0), (0, \pm 1, 0), (0, 0, \pm 1), & i = 2, 3, \cdots, 7, \\ (\pm 1, \pm 1, 0), (\pm 1, 0, \pm 1), (0, \pm 1, \pm 1), & i = 8, 9, \cdots, 19, \\ (\pm 1, \pm 1, \pm 1), & i = 20, 21, \cdots, 27. \end{cases} \quad \text{(B.3)}$$

The constants of the local equilibrium distribution function (1.17) are

$$E_i = \begin{cases} \frac{8}{27}, & i = 1, \\ \frac{2}{27}, & i = 2, 3, \cdots, 7, \\ \frac{1}{54}, & i = 8, 9, \cdots, 19, \\ \frac{1}{216}, & i = 20, 21, \cdots, 27. \end{cases} \quad \text{(B.4)}$$

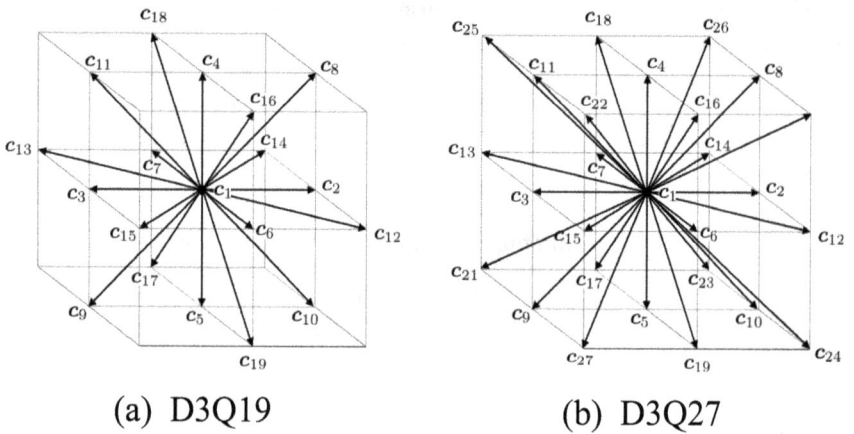

(a) D3Q19   (b) D3Q27

Fig. B   Particle velocities in the (a) D3Q19 and (b) D3Q27 models.

Appendix C

LBM with the MRT Model (MRT-LBM)

To construct the multiple relaxation time (MRT) model [23, 88], we consider variables corresponding to the eigenvectors of the collision scattering matrix A_{ij} in Eq. (1.12), which are created from the moment of the velocity distribution function f_i, and select a different relaxation coefficient for each moment (corresponding to modifying the eigenvalue). The MRT model can compute high Reynolds number flows on a low-resolution lattice with excellent numerical stability. Furthermore, improved MRT models have been proposed, for example, the cascaded collision model [38], the cumulant collision model [40], and the non-orthogonal MRT model [30].

Here, we introduce the LBM that uses the MRT model for the collision term (called the MRT-LBM; see [88]). The following explanation adopts the two-dimensional nine-velocity model (D2Q9 model); MRT-LBMs with the D3Q15, D3Q19, and D3Q27 models are explained in [24, 131].

C.1 Formulation

The MRT-LBM is evolved in time by the following equation:

$$f(x + c_i \Delta x, t + \Delta t) = f(x, t) - M^{-1}\hat{S}[m(x, t) - m^{eq}(x, t)], \quad \text{(C.1)}$$

where f is a column vector of velocity distribution functions f_i:

$$f = (f_1, f_2, \cdots, f_9)^{\mathrm{T}}. \quad \text{(C.2)}$$

In addition, m and m^{eq} are the moments of f_i and f_i^{eq}, respectively, obtained by applying a transformation matrix M to f_i and f_i^{eq}.

$$\boldsymbol{m} = M\boldsymbol{f} = (\rho, e, \epsilon, j_x, q_x, j_y, q_y, p_{xx}, p_{xy})^{\mathrm{T}}, \tag{C.3}$$

$$\boldsymbol{m}^{\mathrm{eq}} = M\boldsymbol{f}^{\mathrm{eq}} = \Big(\rho, -2\rho + 3(j_x^2 + j_y^2)/\rho, \rho - 3(j_x^2 + j_y^2)/\rho,$$

$$j_x, -j_x, j_y, -j_y, (j_x^2 - j_y^2)/\rho, (j_x j_y)/\rho\Big)^{\mathrm{T}}. \tag{C.4}$$

The transformation matrix M is given as follows:

$$M = \begin{bmatrix} 1 & 1 & 1 & 1 & 1 & 1 & 1 & 1 & 1 \\ -4 & -1 & -1 & -1 & -1 & 2 & 2 & 2 & 2 \\ 4 & -2 & -2 & -2 & -2 & 1 & 1 & 1 & 1 \\ 0 & 1 & 0 & -1 & 0 & 1 & -1 & -1 & 1 \\ 0 & -2 & 0 & 2 & 0 & 1 & -1 & -1 & 1 \\ 0 & 0 & 1 & 0 & -1 & 1 & 1 & -1 & -1 \\ 0 & 0 & -2 & 0 & 2 & 1 & 1 & -1 & -1 \\ 0 & 1 & -1 & 1 & -1 & 0 & 0 & 0 & 0 \\ 0 & 0 & 0 & 0 & 0 & 1 & -1 & 1 & -1 \end{bmatrix}. \tag{C.5}$$

Note that the row vectors of M are orthogonal to each other. In Eq. (C.3), ρ is the density, e is the energy, ϵ is the energy squared, $j_x = \rho u_x$ and $j_y = \rho u_y$ are the momenta in the x- and y-directions, respectively, q_x and q_y are the energy fluxes in the x- and y-directions, respectively, p_{xx} is the diagonal component of the stress tensor, and p_{xy} is the off-diagonal component of the stress tensor. Finally, $\hat{S}$ in Eq. (C.1) is a diagonal matrix in which the relaxation parameters $s_1, s_2, \cdots, s_9$ of each moment $(m_1, m_2, \cdots, m_9)^{\mathrm{T}}$ are arranged as follows:

$$\hat{S} = \mathrm{diag}(s_1, s_2, s_3, s_4, s_5, s_6, s_7, s_8, s_9). \tag{C.6}$$

The transformation matrix M, diagonal matrix $\hat{S}$, and collision scattering matrix A in Eq. (1.12) are related through $\hat{S} = -MAM^{-1}$.

In the time evolution equation (C.1), the collision-term computation has different relaxation parameters for each moment, and the velocity distribution functions $(f_1, f_2, \cdots, f_9)^{\mathrm{T}}$ are obtained from the moments $(m_1, m_2, \cdots, m_9)^{\mathrm{T}}$ by applying the inverse transformation M^{-1}. The velocity distribution function is then advected to the adjacent lattice point. This model is called the multiple-relaxation-time (MRT) model because each moment in the collision-term computation can have a different relaxation parameter. Of course, if all relaxation parameters are set to $s_1 = s_2 = \cdots = s_9 = 1/\tau$, Eq. (C.1) becomes the time evolution equation of the SRT-LBM (Eq. (1.14)).

The values of the relaxation parameters are determined as follows. First, the conserved quantities ρ, j_x, and j_y do not need to relax, so we set $s_1 = s_4 = s_6 = 0$. Next, the relaxation parameters of p_{xx} and p_{xy} related to the viscosity coefficient are set to $s_8 = s_9 = 1/\tau$ as in the SRT-LBM. The remaining four relaxation parameters are free and can be arbitrarily determined. To ensure numerical stability, it is recommended that (for example) $s_2 = 1.64$, $s_3 = 1.54$, and $s_5 = s_7 = 1.9$.

In the next section, we will apply the asymptotic analysis (S-expansion) used in Sec. 1.5 to Eq. (C.1). We show that the MRT-LBM also computes incompressible viscous fluids to second-order spatial accuracy. In other words, the SRT-LBM and MRT-LBM differ only by the time evolution equations of $u_\alpha = O((\Delta x)^3)$ and $p = O((\Delta x)^4)$ which include the effect of compressibility. These higher-order time evolution equations also include relaxation parameters other than $s_8 = s_9 = 1/\tau$. The kinematic viscosity coefficient ν and the bulk viscosity coefficient ζ are related to the relaxation parameters of the MRT-LBM as follows:

$$\nu = \frac{1}{3}\left(\frac{1}{s_8} - \frac{1}{2}\right)\Delta x = \frac{1}{3}\left(\tau - \frac{1}{2}\right)\Delta x, \qquad (C.7)$$

$$\zeta = \frac{1}{3}\left(\frac{1}{s_2} - \frac{1}{2}\right)\Delta x. \qquad (C.8)$$

Note that the bulk viscosity coefficient ζ appears in the time evolution equation of $u_\alpha = O((\Delta x)^3)$ (in which many nonlinear terms are added to the Navier–Stokes equations). Also note that the relaxation parameters s_3, s_5, and s_7 cannot be assigned to physical properties, so they have no physical meaning.

As evidenced in Eqs. (C.7) and (C.8), the kinematic viscosity coefficient ν and the bulk viscosity coefficient ζ in the MRT-LBM can be determined with the different relaxation parameters. Therefore, when calculating high Reynolds number flows on a low-resolution lattice (with irreducible Δx),[1] we can set $\zeta \gg \nu$. Thus, as explained in Sec. 2.4, the MRT-LBM has excellent numerical stability. In contrast, the SRT-LBM gives $\zeta = \nu$ (see Table 2.1), so ζ is as small as ν when calculating high Reynolds number flows on a low-resolution lattice. Therefore, the numerical stability is poor. In other words, the better numerical stability of the MRT-LBM than the SRT-LBM in calculations of high Reynolds number flows on a low-resolution lattice is probably explained by the different bulk viscosity coefficients ζ in the two methods [164].

---

[1] From Eq. (C.7) we see that ν must be reduced by setting τ closer to 0.5, but the computation becomes unstable when τ approaches 0.5.

In general, the larger the bulk viscosity coefficient, the better is the numerical stability. Therefore, in recent years, high Reynolds number flows on low-resolution lattices have usually been calculated by the MRT-LBM and improved MRT-LBMs. However, the computational schemes are more complicated than the SRT-LBM, and the computational load increases. Moreover, there is ambiguity in determining many relaxation parameters other than the kinematic viscosity coefficient. Therefore, although the MRT-LBM provides good numerical stability, it must be noted that the relaxation parameters other than $s_8 = s_9 = 1/\tau$ may affect the calculated results within second-order spatial accuracy.

C.2 Asymptotic Analysis (S-Expansion)

Finally, we derive the hydrodynamic equations by applying the asymptotic analysis (S-expansion) of Sec. 1.5 to the MRT-LBM. First, expanding the left-hand side of Eq. (C.1) as a Taylor series around (x, t) up to the order of $(\Delta x)^3$, we have

$$
f(x + c_i \Delta x, t + \Delta t) - f(x, t) = \Delta x \left(I\mathrm{Sh}\frac{\partial}{\partial t} + C_x\frac{\partial}{\partial x} + C_y\frac{\partial}{\partial y} \right) f(x, t)
$$

$$
+ \frac{1}{2}(\Delta x)^2 \left(I\mathrm{Sh}\frac{\partial}{\partial t} + C_x\frac{\partial}{\partial x} + C_y\frac{\partial}{\partial y} \right)^2 f(x, t)
$$

$$
+ \frac{1}{6}(\Delta x)^3 \left(I\mathrm{Sh}\frac{\partial}{\partial t} + C_x\frac{\partial}{\partial x} + C_y\frac{\partial}{\partial y} \right)^3 f(x, t) + \cdots, \tag{C.9}
$$

where I represents the identity matrix, and C_x and C_y represent the following diagonal matrices, respectively:

$$
C_x = \mathrm{diag}(c_{1x}, c_{2x}, c_{3x}, c_{4x}, c_{5x}, c_{6x}, c_{7x}, c_{8x}, c_{9x}), \tag{C.10}
$$

$$
C_y = \mathrm{diag}(c_{1y}, c_{2y}, c_{3y}, c_{4y}, c_{5y}, c_{6y}, c_{7y}, c_{8y}, c_{9y}). \tag{C.11}
$$

Then, multiplying Eq. (C.9) by the transformation matrix M from the left, we get the following equation from Eq. (C.1):

$$
\Delta x \left(I\mathrm{Sh}\frac{\partial}{\partial t} + D_x\frac{\partial}{\partial x} + D_y\frac{\partial}{\partial y} \right) m(x, t)
$$

$$
+ \frac{1}{2}(\Delta x)^2 \left(I\mathrm{Sh}\frac{\partial}{\partial t} + D_x\frac{\partial}{\partial x} + D_y\frac{\partial}{\partial y} \right)^2 m(x, t)
$$

$$
+ \frac{1}{6}(\Delta x)^3 \left(I\mathrm{Sh}\frac{\partial}{\partial t} + D_x\frac{\partial}{\partial x} + D_y\frac{\partial}{\partial y} \right)^3 m(x, t) + \cdots
$$

$$
= -\hat{S}[m(x, t) - m^{\mathrm{eq}}(x, t)], \tag{C.12}
$$

where $D_x = MC_xM^{-1}$ and $D_y = MC_yM^{-1}$. These expressions are explicitly written as follows:

$$D_x = \begin{bmatrix} 0 & 0 & 0 & 1 & 0 & 0 & 0 & 0 & 0 \\ 0 & 0 & 0 & 1 & 1 & 0 & 0 & 0 & 0 \\ 0 & 0 & 0 & 0 & 1 & 0 & 0 & 0 & 0 \\ 2/3 & 1/6 & 0 & 0 & 0 & 0 & 0 & 1/2 & 0 \\ 0 & 1/3 & 1/3 & 0 & 0 & 0 & 0 & -1 & 0 \\ 0 & 0 & 0 & 0 & 0 & 0 & 0 & 0 & 1 \\ 0 & 0 & 0 & 0 & 0 & 0 & 0 & 0 & 1 \\ 0 & 0 & 0 & 1/3 & -1/3 & 0 & 0 & 0 & 0 \\ 0 & 0 & 0 & 0 & 0 & 2/3 & 1/3 & 0 & 0 \end{bmatrix}, \tag{C.13}$$

$$D_y = \begin{bmatrix} 0 & 0 & 0 & 0 & 0 & 1 & 0 & 0 & 0 \\ 0 & 0 & 0 & 0 & 0 & 1 & 1 & 0 & 0 \\ 0 & 0 & 0 & 0 & 0 & 0 & 1 & 0 & 0 \\ 0 & 0 & 0 & 0 & 0 & 0 & 0 & 0 & 1 \\ 0 & 0 & 0 & 0 & 0 & 0 & 0 & 0 & 1 \\ 2/3 & 1/6 & 0 & 0 & 0 & 0 & 0 & -1/2 & 0 \\ 0 & 1/3 & 1/3 & 0 & 0 & 0 & 0 & 1 & 0 \\ 0 & 0 & 0 & 0 & 0 & -1/3 & 1/3 & 0 & 0 \\ 0 & 0 & 0 & 2/3 & 1/3 & 0 & 0 & 0 & 0 \end{bmatrix}. \tag{C.14}$$

We now expand m as follows:

$$m = m^{(0)} + (\Delta x)m^{(1)} + (\Delta x)^2 m^{(2)} + \cdots. \tag{C.15}$$

Accordingly, the macroscopic variables are expanded as follows:

$$\rho = 1 + (\Delta x)\rho^{(1)} + (\Delta x)^2\rho^{(2)} + \cdots, \tag{C.16}$$
$$u_x = (\Delta x)u_x^{(1)} + (\Delta x)^2 u_x^{(2)} + \cdots, \tag{C.17}$$
$$u_y = (\Delta x)u_y^{(1)} + (\Delta x)^2 u_y^{(2)} + \cdots, \tag{C.18}$$

and m^{eq} is expanded as

$$m^{\mathrm{eq}} = m^{\mathrm{eq}(0)} + (\Delta x)m^{\mathrm{eq}(1)} + (\Delta x)^2 m^{\mathrm{eq}(2)} + \cdots, \tag{C.19}$$

where

$$m^{\text{eq}(0)} = \begin{pmatrix} 1 \\ -2 \\ 1 \\ 0 \\ 0 \\ 0 \\ 0 \\ 0 \\ 0 \end{pmatrix}, \tag{C.20}$$

$$m^{\text{eq}(1)} = \begin{pmatrix} \rho^{(1)} \\ -2\rho^{(1)} \\ \rho^{(1)} \\ u_x^{(1)} \\ -u_x^{(1)} \\ u_y^{(1)} \\ -u_y^{(1)} \\ 0 \\ 0 \end{pmatrix}, \tag{C.21}$$

$$m^{\text{eq}(2)} = \begin{pmatrix} \rho^{(2)} \\ -2\rho^{(2)} + 3[(u_x^{(1)})^2 + (u_y^{(1)})^2] \\ \rho^{(2)} - 3[(u_x^{(1)})^2 + (u_y^{(1)})^2] \\ \rho^{(1)} u_x^{(1)} + u_x^{(2)} \\ -(\rho^{(1)} u_x^{(1)} + u_x^{(2)}) \\ \rho^{(1)} u_y^{(1)} + u_y^{(2)} \\ -(\rho^{(1)} u_y^{(1)} + u_y^{(2)}) \\ (u_x^{(1)})^2 - (u_y^{(1)})^2 \\ u_x^{(1)} u_y^{(1)} \end{pmatrix}, \tag{C.22}$$

$$m^{\text{eq}(3)} = \begin{pmatrix} \rho^{(3)} \\ -2\rho^{(3)} + 3[\rho^{(1)}(u_x^{(1)})^2 + \rho^{(1)}(u_y^{(1)})^2] + 6(u_x^{(1)}u_x^{(2)} + u_y^{(1)}u_y^{(2)}) \\ \rho^{(3)} - 3[\rho^{(1)}(u_x^{(1)})^2 + \rho^{(1)}(u_y^{(1)})^2] - 6(u_x^{(1)}u_x^{(2)} + u_y^{(1)}u_y^{(2)}) \\ \rho^{(1)}u_x^{(2)} + \rho^{(2)}u_x^{(1)} + u_x^{(3)} \\ -(\rho^{(1)}u_x^{(2)} + \rho^{(2)}u_x^{(1)} + u_x^{(3)}) \\ \rho^{(1)}u_y^{(2)} + \rho^{(2)}u_y^{(1)} + u_y^{(3)} \\ -(\rho^{(1)}u_y^{(2)} + \rho^{(2)}u_y^{(1)} + u_y^{(3)}) \\ \rho^{(1)}(u_x^{(1)})^2 - \rho^{(1)}(u_y^{(1)})^2 + 2(u_x^{(1)}u_x^{(2)} - u_y^{(1)}u_y^{(2)}) \\ \rho^{(1)}u_x^{(1)}u_y^{(1)} + u_x^{(1)}u_y^{(2)} + u_x^{(2)}u_y^{(1)} \end{pmatrix}.$$

$$\tag{C.23}$$

Substituting these expansions into Eq. (C.12) and collecting terms with the same order of Δx, we obtain the following equations on the diffusive time scale ($\Delta t = \text{Sh}\Delta x = O((\Delta x)^2)$). Specifically, at the zeroth order, we get

$$m^{(0)} = m^{\text{eq}(0)}. \tag{C.24}$$

At the order of Δx, we get

$$\left(D_x \frac{\partial}{\partial x} + D_y \frac{\partial}{\partial y}\right) m^{(0)} = -\hat{S}(m^{(1)} - m^{\text{eq}(1)}). \tag{C.25}$$

At the order of $(\Delta x)^2$, we get

$$\left(D_x \frac{\partial}{\partial x} + D_y \frac{\partial}{\partial y}\right) m^{(1)} = -\hat{S}(m^{(2)} - m^{\text{eq}(2)}). \tag{C.26}$$

At the order of $(\Delta x)^3$, we get

$$I\frac{\text{Sh}}{\Delta x}\frac{\partial m^{(1)}}{\partial t} + \left(D_x \frac{\partial}{\partial x} + D_y \frac{\partial}{\partial y}\right) m^{(2)}$$
$$+ \frac{1}{2}\left(D_x \frac{\partial}{\partial x} + D_y \frac{\partial}{\partial y}\right)^2 m^{(1)} = -\hat{S}(m^{(3)} - m^{\text{eq}(3)}). \tag{C.27}$$

Finally, at the order of $(\Delta x)^4$, we get

$$I\frac{\text{Sh}}{\Delta x}\frac{\partial m^{(2)}}{\partial t} + \left(D_x \frac{\partial}{\partial x} + D_y \frac{\partial}{\partial y}\right) m^{(3)} + \frac{1}{2}\left(D_x \frac{\partial}{\partial x} + D_y \frac{\partial}{\partial y}\right)^2 m^{(2)}$$
$$+ \frac{1}{6}\left(D_x \frac{\partial}{\partial x} + D_y \frac{\partial}{\partial y}\right)^3 m^{(1)} + I\frac{\text{Sh}}{\Delta x}\frac{\partial}{\partial t}\left(D_x \frac{\partial}{\partial x} + D_y \frac{\partial}{\partial y}\right) m^{(1)}$$
$$= -\hat{S}(m^{(4)} - m^{\text{eq}(4)}). \tag{C.28}$$

From Eqs. (C.24) and (C.25), we first obtain

$$\boldsymbol{m}^{(1)} = \boldsymbol{m}^{\text{eq}(1)}. \tag{C.29}$$

Using this result, we obtain Eq. (1.36) from the first component of Eq. (C.26). Similarly, Eq. (1.37) can be obtained from the fourth and sixth components of Eq. (C.26). Proceeding with this analysis and minding the following relationships (which hold for conserved physical quantities): $m_1^{(2)} = m_1^{\text{eq}(2)}$, $m_4^{(2)} = m_4^{\text{eq}(2)}$, $m_6^{(2)} = m_6^{\text{eq}(2)}$ and $m_1^{(3)} = m_1^{\text{eq}(3)}$, $m_4^{(3)} = m_4^{\text{eq}(3)}$, $m_6^{(3)} = m_6^{\text{eq}(3)}$, we obtain Eq. (1.39) from the first component of Eq. (C.27). Also, Eq. (1.40) is obtained from the fourth and sixth components of Eq. (C.27). Finally, Eq. (1.42) is obtained from the first component of Eq. (C.28), and Eq. (1.43) is obtained from the fourth and sixth components of Eq. (C.28).

Note that the relaxation parameters other than $s_8 = s_9 = 1/\tau$ do not appear in the equations obtained up to this order. The relaxation parameter s_2 that determines the bulk viscosity coefficient and the remaining relaxation parameters s_3, s_5, and s_7 appear in the time evolution equation for $u_\alpha^{(3)}$ (in which many nonlinear terms are added to the Navier–Stokes equations) obtained from the equation of the order of $(\Delta x)^5$. That is, to second-order spatial accuracy, the hydrodynamic equations of the MRT-LBM are exactly those of the SRT-LBM. However, as mentioned above, the relaxation parameters other than $s_8 = s_9 = 1/\tau$ may affect the calculated results within second-order spatial accuracy in some cases (for example, when the kinematic viscosity coefficient ν is very small and the bulk viscosity coefficient ζ is large, or when the relaxation parameters s_3, s_5, and s_7 are not given appropriately). In the Chapman–Enskog expansion, the macroscopic variables are not expanded into power series of Δx, so the above situation is not clear. The asymptotic analysis (S-expansion) clarifies the relationship between the orders of the macroscopic variables and the relaxation parameters.

Appendix D

Summation Formulae of Particle Velocities c_i

Below we give the summation formulae of the particle velocities c_i. The following formulae hold in the D2Q9 model ($N = 9$), the D3Q15 model ($N = 15$), the D3Q19 model ($N = 19$), and the D3Q27 model ($N = 27$):

$$\sum_{i=1}^{N} E_i = 1, \tag{D.1}$$

$$\sum_{i=1}^{N} E_i c_{i\alpha} = 0, \tag{D.2}$$

$$\sum_{i=1}^{N} E_i c_{i\alpha} c_{i\beta} = \frac{1}{3}\delta_{\alpha\beta}, \tag{D.3}$$

$$\sum_{i=1}^{N} E_i c_{i\alpha} c_{i\beta} c_{i\gamma} = 0, \tag{D.4}$$

$$\sum_{i=1}^{N} E_i c_{i\alpha} c_{i\beta} c_{i\gamma} c_{i\delta} = \frac{1}{9}(\delta_{\alpha\beta}\delta_{\gamma\delta} + \delta_{\alpha\gamma}\delta_{\beta\delta} + \delta_{\alpha\delta}\delta_{\beta\gamma}), \tag{D.5}$$

$$\sum_{i=1}^{N} E_i c_{i\alpha} c_{i\beta} c_{i\gamma} c_{i\delta} c_{i\epsilon} = 0, \tag{D.6}$$

where N is the number of particle velocities.

Appendix E

Lattice Units

In the LBM, LKS, and improved LKS, dimensionless variables called lattice units with a characteristic length of $\hat{L} = \Delta\hat{x}$ are often used because they are convenient for coding programs of computational schemes. In this case, as $\Delta x_{(lu)} = \Delta\hat{x}/\hat{L} = \Delta\hat{x}/\Delta\hat{x} = 1$, $\Delta t_{(lu)}$ becomes $\Delta t_{(lu)} = \Delta\hat{t}/(\hat{L}/\hat{U}) = (\Delta\hat{x}/\hat{c})/(\Delta\hat{x}/\hat{U}) = \hat{U}/\hat{c} = $ Sh on the diffusive time scale and $\Delta t_{(lu)} = \Delta\hat{t}/(\hat{L}/\hat{c}) = (\Delta\hat{x}/\hat{c})/(\Delta\hat{x}/\hat{c}) = 1$ on the acoustic time scale.

The dimensionless variables in Appendix A are related to the dimensionless variables in lattice units as follows:

$$\begin{cases} \mu = \mu_{(lu)}\Delta x, \ \nu = \nu_{(lu)}\Delta x, \ \sigma = \sigma_{(lu)}\Delta x, \\ \alpha = \alpha_{(lu)}\Delta x, \ g\Delta x = g_{(lu)}, \ M_\phi = M_{\phi(lu)}\Delta x. \end{cases} \tag{E.1}$$

Below, we explain how we determine the dimensionless parameters governing a phenomenon.

(1) Reynolds number Re

For example, suppose that in a computational program coded in lattice units, the flow velocity is $V = 0.01$, the width is $D_{(lu)} = 100$, and the kinematic viscosity coefficient is $\nu_{(lu)} = 0.01$. The Reynolds number Re is then given by

$$\text{Re} = \frac{VD_{(lu)}}{\nu_{(lu)}} = \frac{0.01 \times 100}{0.01} = 100.$$

Now suppose that we use the dimensionless variables in Appendix A. The flow velocity is $V = 0.01$, the width is $D = 100\Delta x$, and the kinematic viscosity coefficient is $\nu = 0.01\Delta x$. The Reynolds number Re is then given by

$$\text{Re} = \frac{VD}{\nu} = \frac{0.01 \times 100\Delta x}{0.01\Delta x} = 100.$$

Obviously, both units give the same Re.

(2) Weber number We

In lattice units with density $\rho_L = 100$, flow velocity $V = 0.01$, width $D_{(lu)} = 100$, and interfacial tension $\sigma_{(lu)} = 0.01$, we get

$$\text{We} = \frac{\rho_L V^2 D_{(lu)}}{\sigma_{(lu)}} = \frac{100 \times (0.01)^2 \times 100}{0.01} = 100.$$

In terms of the dimensionless variables given in Appendix A, setting the density $\rho_L = 100$, flow velocity $V = 0.01$, width $D = 100\Delta x$, and interfacial tension $\sigma = 0.01\Delta x$ gives

$$\text{We} = \frac{\rho_L V^2 D}{\sigma} = \frac{100 \times (0.01)^2 \times 100\Delta x}{0.01\Delta x} = 100.$$

(3) Froude number Fr

In lattice units with flow velocity $V = 0.01$, width $D_{(lu)} = 100$, and gravitational acceleration $g_{(lu)} = 10^{-6}$, we get

$$\text{Fr} = \frac{V}{\sqrt{g_{(lu)} D_{(lu)}}} = \frac{0.01}{\sqrt{10^{-6} \times 100}} = 1.$$

In terms of the dimensionless variables in Appendix A, setting the flow velocity $V = 0.01$, width $D = 100\Delta x$, and gravitational acceleration $g\Delta x = 10^{-6}$ gives

$$\text{Fr} = \frac{V}{\sqrt{gD}} = \frac{0.01}{\sqrt{10^{-6}/\Delta x \times 100\Delta x}} = 1.$$

(4) Rayleigh number Ra

Finally, in lattice units with width $D_{(lu)} = 100$, kinematic viscosity coefficient $\nu_{(lu)} = 0.01$, temperature conductivity coefficient $\alpha_{(lu)} = 0.01$, gravitational acceleration $g_{(lu)} = 10^{-6}$, coefficient of thermal expansion $\beta = 0.1$, and temperature difference $\Delta T = 1$, we get

$$\text{Ra} = \frac{g_{(lu)} \beta (\Delta T) D_{(lu)}^3}{\nu_{(lu)} \alpha_{(lu)}} = \frac{10^{-6} \times 0.1 \times 1 \times (100)^3}{0.01 \times 0.01} = 1000.$$

In terms of the dimensionless variables in Appendix A, setting the width $D = 100\Delta x$, kinematic viscosity coefficient $\nu = 0.01\Delta x$, temperature conductivity coefficient $\alpha = 0.01\Delta x$, gravitational acceleration $g\Delta x = 10^{-6}$, coefficient of thermal expansion $\beta = 0.1$, and temperature difference $\Delta T = 1$ gives

$$\text{Ra} = \frac{g\beta(\Delta T)D^3}{\nu\alpha} = \frac{10^{-6}/\Delta x \times 0.1 \times 1 \times (100\Delta x)^3}{0.01\Delta x \times 0.01\Delta x} = 1000.$$

(5) Capillary number Ca, Ohnesorge number Oh, and Grashof number Gr

The other dimensionless parameters have the following relationships with the above dimensionless parameters:

$$\text{Ca} = \text{We/Re}, \tag{E.2}$$
$$\text{Oh} = \sqrt{\text{We}}/\text{Re}, \tag{E.3}$$
$$\text{Gr} = \text{Ra/Pr}, \tag{E.4}$$

where $\text{Pr} = \nu/\alpha$.

Appendix F

Program Examples

In this Appendix, we provide two basic program examples:

- flow past a circular cylinder;
- a stationary droplet.

These programming examples can be downloaded from the publisher's web page (https://www.worldscientific.com/worldscibooks/10.1142/12375#t=suppl). In the first example, a circular cylinder is placed in a two-dimensional Poiseuille flow. The circular cylinder is configured using the improved bounce-back condition or the immersed boundary method. In the second example, a single liquid droplet is placed in a gas (the liquid-to-gas density ratio is 800), and a pressure difference between the liquid and gas is induced by surface tension.

Both examples are coded in Fortran 90 and C++ with the lattice units shown in Appendix E. The algorithms are shown in the text. These examples are coded as simply as possible, and are therefore inefficient in terms of computational speed. They are intended as learning tools for beginners, along with researchers and engineers who are interested in the LBM computation. It is hoped that users of these programs will gain a better understanding of the LBM.

To visualize the flow field, we recommend open source software such as gnuplot (http://www.gnuplot.info/) and Paraview (https://www.paraview.org/). These URLs are currently available in 2021. In these program examples, data files for gnuplot are output. The usages and data formats of the recommended software packages are not discussed in this Appendix.

F.1 Flow Past a Circular Cylinder

This program example consists of three source codes in Fortran 90:

- a main program named `Poiseuille_2D.f90`;
- subroutine A named `circular_IBB.f90`;
- subroutine B named `circular_IBM.f90`,

or a single source code in C++:

- a main program named `cylinder_in_2DPoiseulle.cpp`.

The main program simulates a two-dimensional Poiseuille flow with the bounce-back condition applied at the upper and lower boundaries of the computational domain. As the inlet and outlet conditions, a pressure difference is imposed at the left and right boundaries. The flow past a circular cylinder is simulated by calling either of the above subroutines in the main program. In subroutines A and B (corresponding to the subroutines `circular_IBB()` and `circular_IBM()`, respectively, in C++ code), the cylinder is configured using the improved bounce-back condition and the immersed boundary method, respectively. If neither of these subroutines is called, this program merely simulates a two-dimensional Poiseuille flow. Note that this example uses the incompressible local equilibrium distribution function given by Eq. (1.58). The algorithms of the main program, subroutine A, and subroutine B are described in Sec. 1.12, (**3**) of Sec. 1.11, and Sec. 3.7, respectively.

In the default settings, the computational domain is divided into a 200×200 square lattice. The diameter of the cylinder is set to $D_{(lu)} = 50$, and the center of the cylinder is located at the center of the domain. The characteristic flow speed U is defined as the mean flow speed over the cross section when the cylinder is absent, and the pressure difference between the left and right boundaries is adjusted to obtain $U = 0.03$. Note that $U \ll c_s$ is set as required by the LBM. Meanwhile, the kinematic viscosity coefficient $\nu_{(lu)}$ is set to obtain $\mathrm{Re} = U H_{(lu)}/\nu_{(lu)} = 100$ at the mean flow speed $U = 0.03$ and wall spacing $H_{(lu)} = 200$. The relaxation time τ is calculated from the kinematic viscosity $\nu_{(lu)}$. All of these computational parameters are written directly in the source codes. If you want to change these settings, you can appropriately rewrite the corresponding part of the source codes.

This program creates output files of the computational parameters, the time variation of the mean flow speed on the cross section, the transient

Reynolds number, and the flow and pressure distributions over the entire computational domain. The flow field can be visualized by loading the data files of the velocity and pressure distributions into gnuplot. If you want other output files, you may change the source codes.

F.2 Stationary Droplet

This program example consists of a single source code:

- a main program named `laplace.f90` or `laplace.cpp`.

The order parameter ϕ is computed by the Cahn–Hilliard equation (described in Sec. 4.3), and the flow velocity and pressure are computed by the improved LKS (described in Sec. 4.7). The algorithm is described in Sec. 4.8. Periodic boundary conditions are imposed on all boundaries of the computational domain.

In the default settings, the computational domain is divided into a $96 \times 96 \times 96$ cubic lattice. The initial diameter of the droplet is $D_{(lu)} = 40$, and the droplet center is located at the center of the domain. The density ratio and viscosity ratio are set to $\rho_L/\rho_G = 800$ and $\mu_L/\mu_G = 50$, respectively, representing an air–water two-phase system. In addition, the surface tension is set to $\sigma_{(lu)} = 1.2 \times 10^{-2}$. All of these computational parameters are written directly in the source code. To change these settings, you can appropriately rewrite the corresponding part of the source code.

This program example creates output files of the computational parameters, the deviation of the gas–liquid pressure difference from the analytical solution, the time variations of the volume and diameter of the droplet, and the distributions of the order parameter ϕ, density ρ, pressure p, and velocity u over the central cross section of the computational domain. If you want other output files, you may change the source code.

As an exercise, readers can modify this program example to simulate droplet collisions.

Afterword

This book summarizes the fundamentals and applications of the LBM based on research results for 27 years after the first author returned to Kyoto University from a research laboratory of the industry in 1994. Initially, I collaborated with the second author on Chap. 1 and some parts of Chap. 2. At that time, we discussed whether our LBM work should continue, as the research seemed to be deadlocked. This discussion has become a nostalgic memory to us both. After that, I proceeded to Chaps. 2 and 3 with the third author. The expansion of the LBM to the IB-LBM created opportunities for interesting research on flapping flight. The contents of Chap. 4 were developed alongside the work of many students who have occupied the laboratory since around 2000. Unlike the contents of the previous chapters, these contents emerged only after repeated trial and error. I give many thanks to the students whose efforts enabled the results of milk crown simulations to be published in this book. Solving this difficult problem has been a long-term goal of computational science.

Looking back over the past 42 years, numerical computations of incompressible viscous fluids have been my main interest since graduating with a Masters' degree in 1979. Instead of directly computing the Navier–Stokes equations, I have attempted indirect numerical solutions to these equations. During the first eight years, I investigated the discrete vortex method, which treats separated flow within the framework of potential flows of an inviscid fluid, and applied it to numerical computations of separated flows around bluff bodies. During the following seven years (including two years (1988–1990) as a visiting associate researcher at Caltech), I worked on numerical computations of molecular gas flows. The Caltech experience and the research conducted during my Masters' program provided useful groundwork for my LBM research. Over the next 27 years, I applied the LBM to various fields as mentioned above.

Throughout my 42-year research career, my aspiration has been "computing complex phenomena by a simple method (framework)." One might ask 'Is it easier to compute the Navier–Stokes equations directly?' At first glance, the obvious answer is "yes," but the handling of nonlinear terms and solving the Poisson equation of pressure present many numerical challenges in complex phenomena. I hope that readers will understand that the LBM greatly simplifies the computation of such complex phenomena.

Finally, I would like to thank Yoshio Sone and Kazuo Aoki, Emeritus Professors of Kyoto University, for their guidance (sometimes severely and sometimes warmly) since my Masters' course research. In addition, I thank the graduates in the laboratories of the Departments of Chemical Engineering and of Aeronautics and Astronautics, Graduate School of Engineering, Kyoto University, who participated in the LBM research related to this book and in the annual OB sandlot baseball. I also thank my family for creating an atmosphere where I can work freely and enjoyably.

October, 2021
Takaji Inamuro

Bibliography

It is not possible to list a large number of papers related to the LBM. We therefore list only the literature cited in this book. Please forgive the authors if important literature has been omitted because it is outside our knowledge scope.

[1] T. Abe, Derivation of the lattice Boltzmann method by means of the discrete ordinate method for the Boltzmann equation, J. Comput. Phys. **131** (1997), 241–246.

[2] C. K. Aidun and J. R. Clausen, Lattice-Boltzmann method for complex flows, Annu. Rev. Fluid Mech. **42** (2010), 439–472.

[3] S. M. Allen and J. W. Cahn, Mechanisms of phase transformation within the miscibility gap of Fe-rich Fe-Al alloys, Acta Metall. **24** (1976), 425–437.

[4] S. Ansumali and I. V. Karlin, Stabilization of the lattice Boltzmann method by the H theorem: A numerical test, Phys. Rev. E **62** (2000), 7999–8003.

[5] P. Asinari, T. Ohwada, E. Chiavazzo, and A. F. Di Rienzo, Link-wise artificial compressibility method, J. Comput. Phys. **231** (2012), 5109–5143.

[6] A. Azuma and T. Watanabe, Flight performance of a dragonfly, J. Exp. Biol. **137** (1988), 221–252.

[7] P. L. Bhatnagar, E. P. Gross, and M. Krook, A model for collision processes in gases. I. Small amplitude processes in charged and neutral one-component systems, Phys. Rev. **94** (1954), 511–525.

[8] R. B. Bird, W. E. Stewart, and E. N. Lightfoot, *Transport Phenomena*, Wiley (1960), 199.

[9] J. U. Brackbill, D. B. Kothe, and C. Zemach, A continuum method for modeling surface tension, J. Comput. Phys. **100** (1992), 335–354.

[10] A. J. Briant, P. Papatzacos, and J. M. Yeomans, Lattice Boltzmann simulations of contact line motion in a liquid-gas system, Phil. Trans. R. Soc. Lond. A **360** (2002), 485–495.

[11] J. M. Buick, C. A. Greated, and D. M. Campbell, Lattice BGK simulation of sound waves, Europhys. Lett. **43** (1998), 235–240.

[12] J. W. Cahn and J. E. Hilliard, Free energy of a nonuniform system. I. Interfacial free energy, J. Chem. Phys. **28** (1958), 258–267.

[13] J. W. Cahn and J. E. Hilliard, Free energy of a nonuniform system. III. Nucleation in a two-component incompressible Fluid, J. Chem. Phys. **31** (1959), 688–699.

[14] S. Chen, H. Chen, D. Martinez, and W. H. Mattaeus, Lattice Boltzmann model for simulation of magnetohydrodynamics, Phys. Rev. Lett. **67** (1991), 3776–3779.

[15] S. Chen and G. D. Doolen, Lattice Boltzmann method for fluid flows, Annu. Rev. Fluid Mech. **30** (1998), 329–364.

[16] L. Chen, Q. Kang, Y. Mu, Y.-L. He, and W.-Q. Tao, A critical review of the pseudopotential multiphase lattice Boltzmann model: Methods and applications, Int. J. Heat Mass Transfer **76** (2014), 210–236.

[17] P. H. Chiu and Y. T. Lin, A conservative phase field method for solving incompressible two-phase flows, J. Comput. Phys. **230** (2011), 185–204.

[18] A. J. Chorin, Numerical solution of the Navier-Stokes equations, Math. Comp. **22** (1968), 745–762.

[19] S. Y. Chou and D. Baganoff, Kinetic flux-vector splitting for the Navier-Stokes equations, J. Comput. Phys. **130** (1997), 217–230.

[20] R. M. Clever and F. H. Busse, Transition to time-dependent convection, J. Fluid Mech. **65** (1974), 625–645.

[21] R. G. Cox, The deformation of a drop in a general time-dependent fluid flow, J. Fluid Mech. **37** (1969), 601–623.

[22] R. A. de Bruijn, Deformation and breakup of drops in simple shear flows, Ph.D. Thesis, Technical University at Eindhoven (1989).

[23] D. d'Humières, Generalized lattice-Boltzmann equations, in: D. Shizgal, D.P. Weaver (Eds.), *Rarefied Gas Dynamics: Theory and Applications*, Progress in Astronautics and Aeronautics, 159, AIAA, Washington, DC (1992), 450–458.

[24] D. d'Humières, I. Ginzburg, M. Krafczyk, P. Lallemand, and L.-S. Luo, Multiple-relaxation-time lattice Boltzmann models in three dimensions, Phil. Trans. R. Soc. Lond. A **360** (2002), 437–451.

[25] P. J. Dellar, Bulk and shear viscosities in lattice Boltzmann equations, Phys. Rev. E **64** (2001), 031203 (11pp).

[26] R. Dudley, Biomechanics of flight in neotropical butterflies: morphometrics and kinematics, J. Exp. Biol. **150** (1990), 37–53.

[27] S. Ergun, Fluid flow through packed columns, Chem. Eng. Prog. **48** (1952), 89–94.

[28] A. Eshghinejadfard and D. Thévenin, Numerical simulation of heat transfer in particulate flows using a thermal immersed boundary lattice Boltzmann method, Int. J. Heat Fluid Flow **60** (2016), 31–46.

[29] A. Fakhari, D. Bolster, and L.-S. Luo, A weighted multiple-relaxation-time lattice Boltzmann method for multiphase flows and its application to partial coalescence cascades, J. Comput. Phys. **341** (2017), 22–43.

[30] L. Fei, K. H. Luo, and Q. Li, Three-dimensional cascaded lattice Boltzmann method: Improved implementation and consistent forcing scheme, Phys. Rev. E **97** (2018), 053309 (12pp).

[31] L. Fei, J. Du, K. H. Luo, S. Succi, M. Lauricella, A. Montessori, and Q. Wang, Modeling realistic multiphase flows using a non-orthogonal multiple-relaxation-time lattice Boltzmann method, Phys. Fluids **31** (2019), 042105 (15pp).

[32] Z.-G. Feng and E. E. Michaelides, The immersed boundary–lattice Boltzmann method for solving fluid–particles interaction problems, J. Comput. Phys. **195** (2004), 602–628.

[33] Z.-G. Feng and E. E. Michaelides, Robust treatment of no-slip boundary condition and velocity updating for the lattice-Boltzmann simulation of particulate flows, Comput. Fluids **38** (2009), 370–381.

[34] J. H. Ferziger and M. Perić, *Computational Methods for Fluid Dynamics*, Springer-Verlag (1996), 202.

[35] O. Filippova and D. Hänel, Grid refinement for lattice-BGK models, J. Comput. Phys. **147** (1998), 219–228.

[36] G. Finotello, J. T. Padding, N. G. Denn, A. Jongsma, F. Innings, and J. A. M. Kuipers, Effect of viscosity on droplet-droplet collisional interaction, Phys. Fluids **29** (2017), 067102 (13pp).

[37] M. Fuchiwaki and K. Tanaka, Three-dimensional vortex structure in the wake of a free-flying butterfly, Trans. JSME **82** (2016), 15-00425 (12pp) (in Japanese).

[38] M. Geier, A. Greiner, and J. G. Korvink, Cascaded digital lattice Boltzmann automata for high Reynolds number flow, Phys. Rev. E **73** (2006), 066705 (10pp).

[39] M. Geier, A. Fakhari, and T. Lee, Conservative phase-field lattice Boltzmann model for interface tracking equation, Phys. Rev. E **91** (2015), 063309 (11pp).

[40] M. Geier, M. Schönherr, A. Pasquali, and M. Krafczyk, The cumulant lattice Boltzmann equation in three dimensions: Theory and validation, Comput. Math. Appl. **70** (2015), 507–547.

[41] A. K. Gunstensen, D. H. Rothman, S. Zaleski, and G. Zanetti, Lattice Boltzmann model of immiscible fluids, Phys. Rev. A **43** (1991), 4320–4327.

[42] Z. Guo, C. Zheng, and B. Shi, Discrete lattice effects on the forcing term in the lattice Boltzmann method, Phys. Rev. E **65** (2002), 046308 (6pp).

[43] X. He, Q. Zou, L.-S. Luo, and M. Dembo, Analytic solutions of simple flows and analysis of nonslip boundary conditions for the lattice Boltzmann BGK model, J. Stat. Phys. **87** (1997), 115–136.

[44] X. He and L.-S. Luo, Lattice Boltzmann model for the incompressible Navier–Stokes equation, J. Stat. Phys. **88** (1997), 927–944.

[45] X. He and L.-S. Luo, A priori derivation of the lattice Boltzmann equation, Phys. Rev. E **55** (1997), 6333–6336.

[46] X. He, S. Chen, and G. D. Doolen, A novel thermal model for the lattice Boltzmann method in incompressible limit, J. Comput. Phys. **146** (1998), 282–300.

[47] X. He, S. Chen, and R. Zhang, A lattice Boltzmann scheme for incompressible multiphase flow and its application in simulation of Rayleigh-Taylor instability, J. Comput. Phys. **152** (1999), 642–663.

[48] X. He, G. D. Doolen, and T. Clark, Comparison of the lattice Boltzmann method and the artificial compressibility method for Navier-Stokes equations, J. Comput. Phys. **179** (2002) 439–451.

[49] F. J. Higuera and J. Jiménez, Boltzmann approach to lattice gas simulations, Europhys. Lett. **9** (1989), 663–668.

[50] H. Hino and T. Inamuro, Turning flight simulations of a dragonfly-like flapping wing-body model by the immersed boundary-lattice Boltzmann method, Fluid Dyn. Res. **50** (2018), 065501 (18pp).

[51] K. Hirohashi and T. Inamuro, Hovering and targeting flight simulations of a dragonfly-like flapping wing-body model by the immersed boundary-lattice Boltzmann method, Fluid Dyn. Res. **49** (2017), 045502 (16pp).

[52] M. Hortmann, M. Perić, and G. Scheuerer, Finite volume multigrid prediction of laminar natural convection: Bench-mark solutions, Int. J. Numer. Methods Fluids **11** (1990), 189–207.

[53] S. A. Hosseini, C. Coreixas, N. Darabiha, and D. Thévenin, Stability of the lattice kinetic scheme and choice of the free relaxation parameter, Phys. Rev. E **99** (2019), 063305 (14pp).

[54] H. Huang, J.-J. Huang, X.-Y. Lu, and M. C. Sukop, On simulations of high-density ratio flows using color-gradient multiphase lattice Boltzmann models, Int. J. Mod. Phys. C **24** (2013), 1350021 (19pp).

[55] J. C. R. Hunt, A. A Wray, and P. Moin, Eddies, streams, and convergence zones in turbulent flows, Proc. of the Summer Program (1988), 193–208.

[56] T. Inamuro and B. Sturtevant, Numerical study of discrete-velocity gases, Phys. Fluids A **2** (1990), 2196–2203.

[57] T. Inamuro, M. Yoshino, and F. Ogino, A non-slip boundary condition for lattice Boltzmann simulations, Phys. Fluids **7** (1995), 2928–2930; Erratum: **8** (1996), 1124.

[58] T. Inamuro, M. Yoshino, and F. Ogino, Accuracy of the lattice Boltzmann method for small Knudsen number with finite Reynolds number, Phys. Fluids **9** (1997), 3535–3542.

[59] T. Inamuro, M. Yoshino, and F. Ogino, Lattice Boltzmann simulation of flows in a three-dimensional porous structure, Int. J. Numer. Methods Fluids **29** (1999), 737–748.

[60] T. Inamuro, M. Yoshino, and F. Ogino, Numerical analysis of unsteady flows in a three-dimensional porous structure, Kagaku Kogaku Ronbunshu **25** (1999), 979–986 (in Japanese).

[61] T. Inamuro, Numerical analysis of complex flows by the lattice Boltzmann method, Nagare **18** (1999), 31–36 (in Japanese).

[62] T. Inamuro, K. Maeba, and F. Ogino, Flow between parallel walls containing the lines of neutrally buoyant circular cylinders, Int. J. Multiphase Flow **26** (2000), 1981–2004.

[63] T. Inamuro, N. Konishi, and F. Ogino, A Galilean invariant model of the lattice Boltzmann method for multiphase fluid flows using free-energy approach, Comput. Phys. Commun. **129** (2000), 32–45.

[64] T. Inamuro, T. Miyahara, and F. Ogino, Lattice Boltzmann simulations of drop deformation and breakup in a simple shear flow, *Computational Fluid Dynamics 2000*, 499–504, Springer-Verlag (2001).

[65] T. Inamuro, Lattice Boltzmann method –New fluid simulation method–, Busei Kenkyu **77** (2001), 197–232 (in Japanese).

[66] T. Inamuro, A lattice kinetic scheme for incompressible viscous flows with heat transfer, Phil. Trans. R. Soc. Lond. A **360** (2002), 477–484.

[67] T. Inamuro, M. Yoshino, H. Inoue, R. Mizuno, and F. Ogino, A lattice Boltzmann method for a binary miscible fluid mixture and its application to a heat-transfer problem, J. Comput. Phys. **179** (2002), 201–215.

[68] T. Inamuro, R. Tomita, and F. Ogino, Lattice Boltzmann simulations of drop deformation and breakup in shear flows, Int.J. Mod. Phys. B **17** (2003), 21–26.

[69] T. Inamuro, T. Ogata, S. Tajima, and N. Konishi, A lattice Boltzmann method for incompressible two-phase flows with large density differences, J. Comput. Phys. **198** (2004), 628–644.

[70] T. Inamuro, T. Ogata, and F. Ogino, Numerical simulation of bubble flows by the lattice Boltzmann method, Future Gener. Comput. Syst. **20** (2004), 959–964.

[71] T. Inamuro, S. Tajima, and F. Ogino, Lattice Boltzmann simulation of droplet collision dynamics, Int. J. Heat Mass Transfer **47** (2004), 4649–4657.

[72] T. Inamuro, Lattice Boltzmann methods for viscous fluid flows and for two-phase fluid flows, Fluid Dyn. Res. **38** (2006), 641–659.

[73] T. Inamuro, Lattice Boltzmann methods for moving boundary flows, Fluid Dyn. Res. **44** (2012), 024001 (21pp).

[74] T. Inamuro, T. Yokoyama, K. Tanaka, and M. Taniguchi, An improved lattice Boltzmann method for incompressible two-phase flows with large density differences, Comput. Fluids **137** (2016), 55–69.

[75] T. Inamuro, Recent trend of research on two-phase lattice Boltzmann method, J. HTSJ **55** (2016), 8–13 (in Japanese).

[76] T. Inamuro, T. Echizen, and F. Horai, Validation of an improved lattice Boltzmann method for incompressible two-phase flows, Comput. Fluids **175** (2018), 83–90.

[77] M. Junk and S. V. Rao, A new discrete velocity method for Navier-Stokes equations, J. Comput. Phys. **155** (1999), 178–198.

[78] M. Junk, A. Klar, and L.-S. Luo, Asymptotic analysis of the lattice Boltzmann equation, J. Comput. Phys. **210** (2005), 676–704.

[79] M. Junk and Z. Yang, Asymptotic analysis of lattice Boltzmann boundary conditions, J. Stat. Phys. **121** (2005), 3–35.

[80] S. K. Kang and Y. A. Hassan, A comparative study of direct-forcing immersed boundary–lattice Boltzmann methods for stationary complex boundaries, Int. J. Numer. Methods Fluids **66** (2011), 1132–1158.

[81] S. Karni, Hybrid multifluid algorithms, SIAM J. Sci. Comput. **17** (1996), 1019–1039.

[82] Y. Kataoka and T. Inamuro, Numerical simulations of the behaviour of a drop in a square pipe flow using the two-phase lattice Boltzmann method, Phil. Trans. R. Soc. A **369** (2011), 2528–2536.

[83] S. H. Kim, H. Pitsch, and I. D. Boyd, Accuracy of higher-order lattice Boltzmann methods for microscale flows with finite Knudsen numbers, J. Comput. Phys. **227** (2008), 8655–8671.

[84] K. Kobayashi, T. Inamuro, and F. Ogino, Numerical simulation of advancing interface in a micro heterogeneous channel by the lattice Boltzmann method, J. Chem. Eng. Jpn. **39** (2006), 257–266.

[85] A. J. C. Ladd, Numerical simulations of particulate suspensions via a discretized Boltzmann equation. Part 1. Theoretical foundation, J. Fluid Mech. **271** (1994), 285–309.

[86] A. J. C. Ladd, Numerical simulations of particulate suspensions via a discretized Boltzmann equation. Part 2. Numerical results, J. Fluid Mech. **271** (1994), 311–339.

[87] M.-C. Lai and C. S. Peskin, An immersed boundary method with formal second-order accuracy and reduced numerical viscosity, J. Comput. Phys. **160** (2000), 705–719.

[88] P. Lallemand and L.-S. Luo, Theory of the lattice Boltzmann method: Dispersion, dissipation, isotropy, Galilean invariance, and stability, Phys. Rev. E **61** (2000), 6546–6562.

[89] P. Lallemand and L.-S. Luo, Lattice Boltzmann method for moving boundaries, J. Comput. Phys. **184** (2003), 406–421.

[90] J. Latt and B. Chopard, Lattice Boltzmann method with regularized precollision distribution functions, Math. Comput. Simul. **72** (2006), 165–168.

[91] T. Lee and C.-L. Lin, A stable discretization of the lattice Boltzmann equation for simulation of incompressible two-phase flows at high density ratio, J. Comput. Phys. **206** (2005), 16–47.

[92] J. Li, Y. Y. Renardy, and M. Renardy, Numerical simulation of breakup of a viscous drop in simple shear flow through a volume-of-fluid method, Phys. Fluids **12** (2000), 269–282.

[93] M. Liu, Z. Yu, T. Wang, J. Wang, and L.-S. Fan, A modified pseudopotential for a lattice Boltzmann simulation of bubbly flow, Chem. Eng. Sci. **65** (2010), 5615–5623.

[94] D. Lycett-Brown, K. H. Luo, R. Liu, and P. Lv, Binary droplet collision simulations by a multiphase cascaded lattice Boltzmann method, Phys. Fluids **26** (2014), 023303 (26pp).

[95] A. Mazloomi, S. S. Chikatamarla, and I. V. Karlin, Entropic lattice Boltzmann method for multiphase flows, Phys. Rev. Lett. **114** (2015), 174502 (5pp).

[96] G. R. McNamara and G. Zanetti, Use of the Boltzmann equation to simulate lattice-gas automata, Phys. Rev. Lett. **61** (1988), 2332–2335.

[97] K. Minami, K. Suzuki, and T. Inamuro, Free flight simulations of a dragonfly-like flapping wing-body model using the immersed boundary-lattice Boltzmann method, Fluid Dyn. Res. **47** (2015), 015505 (17pp).

[98] M. L. Minion and D. L. Brown, Performance of under-resolved two-dimensional incompressible flow simulations, II, J. Comput. Phys. **138** (1997), 734–765.

[99] R. Mittal and G. Iaccarino, Immersed boundary methods, Annu. Rev. Fluid Mech. **37** (2005), 239–261.

[100] A. Montessori, G. Falcucci, M. La Rocca, S. Ansumali, and S. Succi, Three-dimensional lattice pseudo-potentials for multiphase flow simulations at high density ratios, J. Stat. Phys. **161** (2015), 1404–1419.

[101] T. Ohta, *Mathematics of interface dynamics*, Nippon Hyoron sha (1997) (in Japanese).

[102] T. Ohwada and P. Asinari, Artificial compressibility method revisited: Asymptotic numerical method for incompressible Navier–Stokes equations, J. Comput. Phys. **229** (2010), 1698–1723.

[103] T. Ohwada, P. Asinari, and D. Yabusaki, Artificial compressibility method and lattice Boltzmann method: Similarities and differences, Comput. Math. Appl. **61** (2011), 3461–3474.

[104] H. Otomo, R. Zhang, and H. Chen, Improved phase-field-based lattice Boltzmann models with a filtered collision operator, Int. J. Mod. Phys. C **30** (2019), 1941009 (11pp).

[105] K.-L. Pan, K.-L. Huang, W.-T. Hsieh, and C.-R. Lu, Rotational separation after temporary coalescence in binary droplet collision, Phys. Rev. Fluids **4** (2019), 123602 (20pp).

[106] C. S. Peskin, Flow patterns around heart valves: A numerical method, J. Comput. Phys. **10** (1972), 252–271.

[107] C. S. Peskin, Numerical analysis of blood flow in the heart, J. Comput. Phys. **25** (1977), 220–252.

[108] C. S. Peskin, The immersed boundary method, Acta Numer. **11** (2002), 479–517.

[109] Y. H. Qian, D. d'Humières, and P. Lallemand, Lattice BGK models for Navier-Stokes equation, Europhys. Lett. **17** (1992), 479–484.

[110] J. Qian and C. K. Law, Regimes of coalescence and separation in droplet collision, J. Fluid Mech. **331** (1997), 59–80.

[111] J. D. Ramshaw and V. A. Mousseau, Accelerated artificial compressibility method for steady-state incompressible flow calculations, Comput. Fluids **18** (1990), 361–367.

[112] W. W. Ren, C. Shu, and W. M. Yang, An efficient immersed boundary method for thermal flow problems with heat flux boundary conditions, Int. J. Heat Mass Transfer **64** (2013), 694–705.

[113] M. Rieber and A. Frohn, A numerical study on the mechanism of splashing, Int. J. Heat Fluid Flow **20** (1999), 455–461.

[114] D. H. Rothman and S. Zaleski, *Lattice-Gas Cellular Automata*, Cambridge University Press (1997).

[115] J. S. Rowlinson and B. Widom, *Molecular Theory of Capillarity*, Clarendon, Oxford (1989), 50–68 (Chapter 3).

[116] F. D. Rumscheidt and S. G. Mason, Deformation and burst of fluid drops in shear and hyperbolic flow, J. Colloid Sci. **16** (1961), 238–261.

[117] B. Sakakibara and T. Inamuro, Lattice Boltzmann simulation of collision dynamics of two unequal-size droplets, Int. J. Heat Mass Transfer **51** (2008), 3207–3216.

[118] M. D. Saroka, N. Ashgriz, and M. Movassat, Numerical investigation of head-on binary drop collisions in a dynamically inert environment, J. Appl. Fluid Mech. **5** (2012), 23–37.

[119] N. Satofuka and T. Nishioka, Parallelization of lattice Boltzmann method for incompressible flow computations, Comput. Mech. **23** (1999), 164–171.

[120] T. Seta, Implicit temperature-correction-based immersed-boundary thermal lattice Boltzmann method for the simulation of natural convection, Phys. Rev. E **87** (2013), 063304 (16pp).

[121] T. Seta, R. Rojas, K. Hayashi, and A. Tomiyama, Implicit-correction-based immersed boundary-lattice Boltzmann method with two relaxation times, Phys. Rev. E **89** (2014), 023307 (22pp).

[122] X. Shan and H. Chen, Lattice Boltzmann model for simulating flows with multiple phases and components, Phys. Rev. E **47** (1993), 1815–1820.

[123] Y. P. Sitompul and T. Aoki, A filtered cumulant lattice Boltzmann method for violent two-phase flows, J. Comput. Phys. **390** (2019), 93–120.

[124] G. D. Smith, *Numerical solution of partial differential equations: finite difference methods*, Oxford University Press (1985), 262–266.

[125] Y. Sone, Asymptotic theory of flow of rarefied gas over a smooth boundary II, *Rarefied Gas Dynamics*, ed. D. Dini (Editrice Tecnico Scientifica, Pisa) **2** (1971), 737–749.

[126] Y. Sone and K. Aoki, *Molecular Gas Dynamics*, Asakura Publishing (1994) (in Japanese).

[127] Y. Sone, Notes on kinetic-equation approach of fluid-dynamic equations, Mechanik-KTH 2000:09, Department of Mechanics, Royal Institute of Technology, Stockholm (2000).

[128] J. D. Sterling and S. Chen, Stability analysis of lattice Boltzmann methods, J. Comput. Phys. **123** (1996), 196–206.

[129] G. Strang, *Linear Algebra and Its Applications*, Academic Press (1976), 81.

[130] S. Succi, *The Lattice Boltzmann Equation for Fluid Dynamics and Beyond*, Oxford University Press (2001).

[131] K. Suga, Y. Kuwata, K. Takashima, and R. Chikasue, A D3Q27 multiple-relaxation-time lattice Boltzmann method for turbulent flows, Comput. Math. Appl. **69** (2015), 518–529.

[132] M. Sun and S. L. Lan, A computational study of the aerodynamic forces and power requirements of dragonfly (*Aeschna juncea*) hovering, J. Exp. Biol. **207** (2004), 1887–1901.

[133] K. Suzuki and T. Inamuro, Effect of internal mass in the simulation of a moving body by the immersed boundary method, Comput. Fluids **49** (2011), 173–187.

[134] K. Suzuki and T. Inamuro, A higher-order immersed boundary–lattice Boltzmann method using a smooth velocity field near boundaries, Comput. Fluids **76** (2013), 105–115.

[135] K. Suzuki and T. Inamuro, An improved lattice kinetic scheme for incompressible viscous fluid flows, Int. J. Mod. Phys. C **25** (2014), 1340017 (9pp).

[136] K. Suzuki, K. Minami, and T. Inamuro, Lift and thrust generation by a butterfly-like flapping wing–body model: immersed boundary–lattice Boltzmann simulations, J. Fluid Mech. **767** (2015), 659–695.

[137] K. Suzuki and M. Yoshino, Numerical simulations for aerodynamic performance of a butterfly-like flapping wing–body model with various wing planforms, Commun. Comput. Phys. **23** (2018), 951–979.

[138] K. Suzuki, T. Kawasaki, N. Furumachi, Y. Tai, and M. Yoshino, A thermal immersed boundary–lattice Boltzmann method for moving-boundary flows with Dirichlet and Neumann conditions, Int. J. Heat Mass Transfer **121** (2018), 1099–1117.

[139] K. Suzuki, I. Okada, and M. Yoshino, Effect of wing mass on the free flight of a butterfly-like model using immersed boundary–lattice Boltzmann simulations, J. Fluid Mech. **877** (2019), 614–647.

[140] K. Suzuki, T. Aoki, and M. Yoshino, Effect of chordwise wing flexibility on flapping flight of a butterfly model using immersed-boundary lattice Boltzmann simulations, Phys. Rev. E **100** (2019), 013104 (16pp).

[141] K. Suzuki, T. Inamuro, and M. Yoshino, Asymptotic equivalence of forcing terms in the lattice Boltzmann method within second-order accuracy, Phys. Rev. E **102** (2020), 013308 (11pp).

[142] K. Suzuki, T. Inamuro, A. Nakamura, F. Horai, K.-L. Pan, and M. Yoshino, Simple extended lattice Boltzmann methods for incompressible viscous single-phase and two-phase fluid flows, Phys. Fluids **33** (2021), 037118 (15pp).

[143] M. R. Swift, W. R. Osborn, and J. M. Yeomans, Lattice Boltzmann simulation of nonideal fluids, Phys. Rev. Lett. **75** (1995), 830–833.

[144] M. R. Swift, E. Orlandini, W. R. Osborn, and J. M. Yeomans, Lattice Boltzmann simulations of liquid-gas and binary fluid systems, Phys. Rev. E **54** (1996), 5041–5052.

[145] N. Takada, J. Matsumoto, and S. Matsumoto, A diffuse-interface tracking method for the numerical simulation of motions of a two-phase fluid on a solid surface, J. Comput. Multiphase Flows **6** (2014), 283–298.

[146] A. ten Cate, C. H. Nieuwstad, J. J. Derksen, and H. E. A. Van den Akker, Particle imaging velocimetry experiments and lattice-Boltzmann simulations on a single sphere settling under gravity, Phys. Fluids **14** (2002), 4012–4025.

[147] M. Tsutahara, T. Kataoka, K. Shikata, and N. Takada, New model and scheme for compressible fluids of the finite difference lattice Boltzmann method and direct simulations of aerodynamic sound, Comput. Fluids **37** (2008), 79–89.

[148] M. Tsutahara, *Lattice Boltzmann method / Difference lattice Boltzmann method*, Corona Publishing (2018) (in Japanese).

[149] M. Uhlmann, An immersed boundary method with direct forcing for the simulation of particulate flows, J. Comput. Phys. **209** (2005), 448–476.

[150] E. M. Viggen, Acoustic multipole sources for the lattice Boltzmann method, Phys. Rev. E **87** (2013), 023306 (5pp).

[151] Z. Wang, J. Fan, and K. Luo, Combined multi-direct forcing and immersed boundary method for simulating flows with moving particles, Int. J. Multiphase Flow **34** (2008), 283–302.

[152] Z. Wang, J. Fan, K. Luo, and K. Cen, Immersed boundary method for the simulation of flows with heat transfer, Int. J. Heat Mass Transfer **52** (2009), 4510–4518.

[153] M. Watari and M. Tsutahara, Two-dimensional thermal model of the finite-difference lattice Boltzmann method with high spatial isotropy, Phys. Rev. E **67** (2003), 036306 (7pp).

[154] D. A. Wolf-Gladrow, *Lattice-Gas Cellular Automata and Lattice Boltzmann Models*, Springer-Verlag (2000).

[155] Z. Xia, K. W. Connington, S. Rapaka, P. Yue, J. J. Feng, and S. Chen, Flow patterns in the sedimentation of an elliptical particle, J. Fluid Mech. **625** (2009), 249–272.

[156] W. Xian and T. Aoki, Multi-GPU performance of incompressible flow computation by lattice Boltzmann method on GPU cluster, Parallel Comput. **37** (2011), 521–535.

[157] K. Yamamoto, X. He, and G. D. Doolen, Simulation of combustion field with lattice Boltzmann method, J. Stat. Phys. **107** (2002), 367–383.

[158] A. J. M. Yang, P. D. Fleming III, and J. H. Gibbs, Molecular theory of surface tension, J. Chem. Phys. **64** (1976), 3732–3742.

[159] A. L. Yarin and D. A. Weiss, Impact of drops on solid surfaces: self-similar capillary waves, and splashing as a new type of kinematic discontinuity, J. Fluid Mech. **283** (1995), 141–173.

[160] N. Yokoyama, K. Senda, M. Iima, and N. Hirai, Aerodynamic forces and vortical structures in flapping butterfly's forward flight, Phys. Fluids **25** (2013), 021902 (24pp).

[161] H. Yoshida and M. Nagaoka, Multiple-relaxation-time lattice Boltzmann model for the convection and anisotropic diffusion equation, J. Comput. Phys. **229** (2010), 7774–7795.

[162] M. Yoshino and T. Inamuro, Lattice Boltzmann simulations for flow and heat/mass transfer problems in a three-dimensional porous structure, Int. J. Numer. Methods Fluids **43** (2003), 183–198.

[163] M. Yoshino, J. Sawada, and K. Suzuki, Numerical simulation of head-on collision dynamics of binary droplets with various diameter ratios by the two-phase lattice kinetic scheme, Comput. Fluids **168** (2018), 304–317.

[164] D. Yu, R. Mei, L.-S. Luo, and W. Shyy, Viscous flow computations with the method of lattice Boltzmann equation, Prog. Aerosp. Sci. **39** (2003), 329–367.

[165] P. Yue, C. Zhou, and J. J. Feng, Spontaneous shrinkage of drops and mass conservation in phase-field simulations, J. Comput. Phys. **223** (2007), 1–9.

[166] H. W. Zheng, C. Shu, and Y. T. Chew, A lattice Boltzmann model for multiphase flows with large density ratio, J. Comput. Phys. **218** (2006), 353–371.

[167] Q. Zou, S. Hou, S. Chen, and G. D. Doolen, An improved incompressible lattice Boltzmann model for time-independent flows, J. Stat. Phys. **81** (1995), 35–48.

Index

www.ingramcontent.com/pod-product-compliance
Lightning Source LLC
Chambersburg PA
CBHW050630190326
41458CB00008B/2208